Management for Professionals

More information about this series at http://www.springer.com/series/10101

José Luís Quesado Pinto •
João Carlos O. Matias • Carina Pimentel •
Susana Garrido Azevedo • Kannan Govindan

Just in Time Factory

Implementation Through Lean Manufacturing Tools

Springer

José Luís Quesado Pinto
LATAM Airlines
Sao Paulo, Brazil

João Carlos O. Matias
DEGEIT
University of Aveiro
Aveiro, Portugal

Carina Pimentel
DEGEIT
University of Aveiro
Aveiro, Portugal

Susana Garrido Azevedo
Department of Business and Economics
University of Beira Interior
Covilha, Portugal

Kannan Govindan
Department of Technology and Innovation
University of Southern Denmark
Odense M, Denmark

ISSN 2192-8096 ISSN 2192-810X (electronic)
Management for Professionals
ISBN 978-3-030-08349-6 ISBN 978-3-319-77016-1 (eBook)
https://doi.org/10.1007/978-3-319-77016-1

Printed on acid-free paper

This Springer imprint is published by the registered company Springer International Publishing AG part of Springer Nature.
The registered company address is: Gewerbestrasse 11, 6330 Cham, Switzerland

Foreword

Toyota Production System (TPS), first developed in circa 1950 by Taiichi Ohno, was inspired because of necessity, as the postwar Japanese consumer was hungry for a large variety of automobiles in small quantities rather than mass-produced automobiles of the same type in large quantities, as was typical in those days when US auto companies used to produce automobiles using the Ford Motor Company's principle of mass-producing the same type of automobiles in large quantities. TPS, which evolved into the more popularly known terminologies of just-in-time and Kanban systems, is a predecessor and an inspiration for the more general lean manufacturing. Lean manufacturing is an approach that focuses on eliminating waste by reducing costs in production, operations, and utilization of equipment and labor. Typically, there are seven types of wastes that should be avoided, viz., overproduction, waiting, unnecessary transportation, overprocessing, excess inventory, unnecessary motion, and defects. Waste has a direct impact on costs, which, if unchecked, could lead to as much as 90% of operations cost run-up. Therein lie the benefits of implementing lean manufacturing, which zeroes in and targets the elimination of wastes and leads to significant reduction in operations and materials costs.

In this book, *The Implementation of a Just-In-Time Factory through Lean Manufacturing Tools*, Pinto, Matias, Pimentel, Azevedo, and Govindan offer a new perspective in the field of just-in-time implementation. A key feature of this book is that it provides a holistic and compact view of just-in-time and lean manufacturing. The book is divided into five chapters. The book starts with an overall introduction to the domain of the area covered in this monograph. The second chapter addresses the topic of lean manufacturing and Kaizen in detail, including the foundations, principles, types of wastes, and benefits of the implementation of lean manufacturing as well as the fundamentals and innovations aspects of Kaizen. The topic of the third chapter is just-in-time (JIT). The coverage includes the foundations of the JIT system, JIT as a tool, its relationships with production management and human dimension, its role in TPS structure, and its overall impact. The focus of Chap. 4 is on the lean manufacturing tools including the five "S," the standardized work, single-minute exchange of die (SMED), and the Kanban system. The last chapter comments on the framework and outlook of just-in-time implementation. The authors of the book form a team who may be viewed as a fine and synergic combination of complementary experiences and backgrounds from academia and industry. Together, they have authored/coauthored hundreds of technical papers and are pioneers in the domain of the book.

This book represents a very valuable work and presents practical application of many lean tools. The organization of the monograph is well thought, and the presentation style is rigorous and clear. Subsequently, though information content is very dense and diverse, the book is accessible and its study is scientifically rewarding. The book is a timely work which contains relevant, inspiring, and practical information.

Northeastern University
Boston, MA, USA

Surendra M. Gupta

Acknowledgements

The current study was funded by Fundação para a Ciência e Tecnologia (FCT), under project UID/EMS/00151/2013 C-MAST, with reference POCI-01-0145-FEDER-007718 and CEFAGE-UBI, which has financial support from FCT, Portugal, and FEDER/COMPETE 2020, through grant UID/ECO/04007/2013 (POCI-01-0145-FEDER-007659). This work was also financially supported by the research unit on Governance, Competitiveness and Public Policy (project POCI-01-0145-FEDER-008540), funded by FEDER funds through COMPETE2020—Programa Operacional Competitividade e Internacionalização (POCI)—and by national funds through FCT—Fundação para a Ciência e a Tecnologia.

Introduction to the Content of the Book

Just-in-time (JIT) production has been one of the hottest research areas in operations management since the 1980s. Centered on JIT production, some experts have proposed more integrative production systems combining JIT production with other approaches and methods in management such as the lean manufacturing system. Due to lean manufacturing a new production paradigm emerged, which confronted the deeply recognized mass production system, contributing to a change in the trade-off between productivity, quality, and delivery. It also led to a rethinking on the way that other operations beyond the high-volume repetitive manufacturing environment are managed.

Despite the vast body of literature about JIT, there is still room for improvement in what concerns the way it should be implemented and also the required lean manufacturing tools to support its implementation and enhance the results achieved. So, the main contribution of this book is to highlight the need for a previous application of a set of vital lean manufacturing tools to support the implementation of JIT system.

This book beyond the description of the JIT production system also explores the putting into practice of a set of lean manufacturing preparatory tools for the JIT system implementation. Our intent is to show the benefits of using 5S, Kaizen, standardized work, SMED, Kanban, and Heijunka as elementary tools for a successful establishment and sustenance of JIT. This book will give insights into the processes associated with the application of these tools detailing their main elements and underlining the main advantages that may be reached by companies seeking to implement the JIT system. To a deeper understanding of the lean manufacturing tools deployment, a real case study is used along the book.

The book is organized into five chapters.

In Chap. 1, an introduction to lean manufacturing and just-in-time is present. Beyond the definition of these two concepts it highlights the main benefits associated with their implementation.

Chapter 2 answers the question: What is lean manufacturing? This chapter covers a brief description of its foundations, its main principles and wastes, and also the benefits companies can reach with the implementation of lean manufacturing best practices. The chapter also describes the Kaizen philosophy reflecting the concern with a continuous improvement culture, where strategic areas of generic continuous improvement behaviors are enhanced. The Kaizen philosophy is also presented as an innovation driver.

Chapter 3 presents the foundations of just-in-time with the identification of the main assumptions and requirements that should be observed to its success with a special emphasis on the human dimensions of JIT. The main positive impacts of a successful implementation of JIT are also identified.

In Chap. 4, the main lean manufacturing tools are identified and described. The lean tools explored in this chapter are 5S, standardized work, SMED—single-minute exchange of die, and Kanban System. Examples of practical applications and some worksheets are incorporated to help managers in the implementation issues.

Finally, in Chap. 5, some considerations about the just-in-time and lean manufacturing are drawn.

The targeted audience for this book are practicing professionals, including executives, middle management, and supervisors and also graduate students from higher education schools in the areas of industrial engineering, engineering and industrial management, and management.

Contents

About the Authors

Susana Garrido Azevedo is Associate Professor with Habilitation at the University of Beira Interior (Portugal). Member of the Research units: CEFAGE-UBI (Center for Advanced Studies in Management and Economics) and UNIDEMI (R&D Unit in Mechanical & Industrial Engineering of the Faculty of Sciences and Technology—FCT NOVA). Her research interests include sustainability, supply chain management, lean, green, and logistics. She has more than 300 scientific works published in books, chapters, articles, and conference proceedings. She is a reviewer in many journals and international conferences. She is Editor-in-Chief, associate editor, and editorial board member of several international journals.

Kannan Govindan is a Full Professor and Head of the Center for Sustainable Supply Chain Engineering, University of Southern Denmark, Odense, Denmark. He was awarded the Gold Medal for the best Ph.D. thesis from the former president of India. Also, he was awarded Young Alumni Achiever Award for Excellence in Academic/Research from the governor of Andhra Pradesh and Telangana. He received the International Young Scientists award from the Chinese Academy of Science, 2012–2013. His research interests include reverse logistics, closed-loop supply chain, sustainable circular economy, green supply chain management, and sustainable supply chain management. He has published more than 200 international journal articles in leading journals such as *Nature*, *Omega*, *Journal of the Operational Research Society*, *Journal of Environmental Management*, *Renewable and Sustainable Energy Reviews*, *European Journal of Operational Research*, *Journal of Cleaner Production*, *Computers & Industrial Engineering*, *International Journal of Production Research*, *Ecological Indicators*, *Transportation Research Part E: Logistics and Transportation Review*, *Transportation Research Part D: Transport and Environment*, *Transactions on Systems, Man, and Cybernetics Part A: Systems and Humans*, *European Management Journal*, *Computers and OR*, *International Journal of Production Economics*, and *International Journal of Production Research*. His research has covered operations management, operations research, supply chain management, lean management, circular economy, and environmental and social issues. Many of his papers were selected as the ESI top 1% highly cited papers or 0.1% hot papers or identified and highlighted as the Key Scientific Article contributing to the excellence in engineering and environmental research (Source:

Web of Science). He has held various teaching and research roles in a number of European, Asian, and American universities, including Dalian Maritime University (China), Nankai University (China), Tianjin University (China), Tianjin University of Technology (China), Luangwa University of Science and Technology (Taiwan), São Paulo State University (Brazil), Clark University (USA), and Universidade Federal de Santa Catarina (Brazil). Currently, he is an Editor-in-Chief of *the International Journal of Business Performance and Supply Chain Modelling* (Inderscience) and the *International Journal of Advanced Operations Management* (Inderscience). In addition, he serves as associate editor of the *Journal of Cleaner Production* and has served as a guest editor in journals such as *European Journal of Operations Research*, *Computers and OR*, *Annals of OR*, *Journal of Cleaner Production*, and *International Journal of Production Economics*. Finally, he serves as an Editorial Board Member of several international journals.

Surendra M. Gupta, Ph.D., P.E., is a Professor of Mechanical and Industrial Engineering and the Director of the Laboratory for Responsible Manufacturing, Northeastern University. He received his BE in electronics engineering from Birla Institute of Technology and Science, MBA from Bryant University, and MSIE and Ph.D. in industrial engineering from Purdue University. He is a registered professional engineer in the state of Massachusetts. Dr. Gupta's research interests span the areas of production/manufacturing systems and operations research. He is mostly interested in lean systems, environmentally conscious manufacturing, reverse and closed-loop supply chains, disassembly modeling, and remanufacturing. He has authored or coauthored ten books and over 550 technical papers published in edited books, journals, and international conference proceedings. His publications have received over ten thousand citations from researchers all over the world in journals, proceedings, books, and dissertations. He has traveled to all seven continents, viz., Africa, Antarctica, Asia, Australia, Europe, North America, and South America, and presented his work at international conferences on six continents. Dr. Gupta has taught over 150 courses in such areas as operations research, inventory theory, queuing theory, engineering economy, supply chain management, and production planning and control. Among the many recognitions received, he is the recipient of outstanding research award and outstanding industrial engineering professor award (in recognition of teaching excellence) from Northeastern University as well as a national outstanding doctoral dissertation advisor award.

João C.O. Matias is Full Professor at the Department of Economics, Management, Industrial Engineering and Tourism (DEGEIT), University of Aveiro (UA); Researcher at C-MAST/UBI and GOVCOPP/UA; and member of the Industrial Engineering and Management and Sustainable Energy Research Groups. He is Coordinator of Scientific Area and Director of Ph.D. Course in Industrial Engineering and Management at DEGEIT/UA. His areas of research focus on industrial engineering and management in general and in sustainability, sustainable energy systems, lean, quality, and management systems in particular. He is involved in several research projects, and he is Editor-in-Chief of a scientific journal and member of Editorial Board of several scientific journals, and he has collaborated as reviewer with many journals and also international conferences. He is author or coauthor of more than 250 scientific works published in several international journals and conference proceedings.

Carina Pimentel is an Industrial Engineering Assistant Professor at the Department of Economics, Management, Industrial Engineering and Tourism of the University of Aveiro. She is a researcher at the Decision Support Systems group of the Research Unit on Governance, Competitiveness and Public Policies (GOVCOPP). Her main research interests include the development of solution approaches for supply chain management and operations management decision problems, namely in the production planning and scheduling, lean manufacturing, supply chain strategic decisions, and urban logistics areas.

José Luís Quesado Pinto is Corporate Senior Manager—Services Development & Innovation at LATAM Airlines, South America's leading airline group. He has over 12 years of international experience within industries such as automotive, electronics, packaging, and aviation, always associated with lean manufacturing and operations areas. His main research and work interests are under these fields, being author and coauthor of published books associated with lean as well as with the food industry/HACCP.

1 Introduction to Lean and Just-in-Time Manufacturing

Lean manufacturing has for some time been considered a strategic management philosophy adopted not only by individual companies but also supply chains to reach higher levels of efficiency and be more competitive (Wong & Wang, 2011). It can also contribute to increasing the environmental and economic performance within a supply chain domain. As argued by Wong and Wang (2011), its objective is to make organizations more efficient and responsive by using fewer resources (human, inventory, time, space), contributing at the same time to increasing the quality of products (Shah & Ward, 2007; Womack & Jones, 1996; Womack, Jones, & Roos, 1990).

Deif and ElMaraghy (2014) note that the business environment is characterized by high uncertainty in terms of demand and customer expectations, which represents a challenge to companies since in this context the known Taylor–Ford model is no longer applicable. Production in large batches has been replaced by smaller batches of a high variety of items designed to meet customer expectations using more efficient production processes.

Lean manufacturing is widely used by industries mainly to mitigate and eliminate all kinds of waste and to improve productivity as a way of enhancing the competitiveness of organizations (Chaple, Narkhede, & Akarte, 2014). This philosophy was introduced and tested in the Toyota production system (TPS) in the 1970s, gaining recognition with the publication of the book "*The Machine that Changed the World*" (Katayama & Bennett, 1996). It is mainly based on improving manufacturing flows, customer-driven production, visual management and an ergonomic working environment (Womack & Jones, 2003).

A positive correlation between lean implementation and business performance has been highlighted in numerous researchers, including the extensive research carried out by Losonci and Demeter (2013). According to Achanga (2006), the main benefits associated with the implementation of lean manufacturing derives from the reduction in machine setup times, inventory levels and waste and better control of production through kanban boards. Beyond these aspects, the same author also highlights the importance of companies developing partnerships with their

J. L. Q. Pinto et al., *Just in Time Factory*, Management for Professionals,
https://doi.org/10.1007/978-3-319-77016-1_1

suppliers, optimizing the production mix, investing in quality management programmes, kaizen and the implementation of error-proof systems.

The reduction of inventory levels and various types of waste has become mandatory, ensuring that manufacturing systems are responsive and not wasteful to allow the implementation of the pull production system. In the pull production system, the products are manufactured after the customer has placed an order, thus avoiding by this way the waste associated to the excess of inventory (Achanga, 2006).

Partnerships with suppliers comprise another critical factor for the success of lean manufacturing implementation, requiring that companies work closer with their suppliers to share developments, divide responsibilities and ensure accurate delivery through just-in-time production (Myerson, 2012).

Wyrwicka and Mrugalska (2017) have identified many organizational, human and technological obstacles that represent important barriers to the implementation of lean manufacturing. As examples of organizational obstacles, they consider the dimension of plants, a weak relationship between strategy and operational improvements, high compartmentalization of functional units, organizational culture, etc. Some human obstacles to the implementation of lean manufacturing are the following: negative attitudes of staff, strong resistance to change, lack of management support and commitment, low skill of workers. Various technological obstacles were identified, such as lack of technological infrastructure and insufficient technical resources.

With regard to just-in-time (JIT) production, it is considered to be a production system for making and delivering what is needed, just when it is needed and just in the amount needed (Narusawa & Shook, 2009). More broadly, under JIT, all organizations should be managed with a concern for continuous improvement in which non-value-added activities are identified and removed in order to reduce costs and improve quality and delivery (Chen & Tan, 2011). This system originated at Toyota during the development of the TPS and is typically credited to Mr. Taiichi Ohno (1988). The primary objective of the TPS and JIT was to increase operational excellence by consistently and thoroughly eliminating waste, improving quality and reducing costs, the utilization of resources and the time associated with production and delivery. Although it is relatively easy in theory to understand that JIT contributes to eliminating waste and improving operational excellence, in reality it is quite difficult to implement JIT production as it requires many organizational changes associated with the reengineering of processes and the management of physical and informational flows (Wafa & Yasin, 1998).

Originally, JIT was mainly adopted by industries focused on production processes. However, with a new focus on supply chains, JIT practices have been extended upstream and downstream in supply chains to include purchasing and selling linkages, giving rise to the Total JIT concept (Claycomb, Germain, & Dröge, 1999). Total JIT is considered to comprise an integrated supply chain strategy, incorporating not only production, but also purchasing, selling and information elements (Green, Inman, Birou, & Whitten, 2014).

Successful implementation of JIT in an organization largely depends on front-line employees. Their role is to work on improving of work processes, understand quality measures, solve problems, generate a broader outlook of the production process, ensure the inter-connectivity of workers and decide on basic maintenance practices. They need to understand the condition of machines and equipment and have the ability to meet quality and production requirements. Thus, employee empowerment and involvement represent key elements in maintenance practice and need to be measured as JIT elements for maintenance (Phogat & Gupta, 2017).

The main idea associated with JIT thinking is that production should be "pulled through" rather than "pushed through". This means that the production cycle starts only once a customer has placed an order and stocks are delivered when they are needed, as a consequence requiring much more frequent delivery of stock.

JIT is the key element in what is termed lean production and the concept has been confused with lean manufacturing. However, while the efficiency is the focus of JIT manufacturing (Monden, 1993), lean manufacturing is focused on using efficiency to add value for the customer (Locher, 2008; Ward, 2007). JIT manufacturing can be practiced on its own or as one step in the lean manufacturing process. There are other differences between JIT and lean. Lean is an integrated system that can be used across business departments, including manufacturing, production, marketing and distribution (Hobbs, 2011). Lean creates a JIT process responsible for eliminating the waste of excessive inventory.

References

Achanga, P. (2006). Critical success factors for lean implementation within SMEs. *Journal of Manufacturing Technology Management, 17*(4), 460–471.

Chaple, A. P., Narkhede, B. E., & Akarte, M. M. (2014). Status of implementation of lean manufacturing principles in the context of Indian industry: A literature review. Proceedings from *5th International and 26th All India Manufacturing Technology, Design and Research Conference* (AIMTDR 2014) December 12th–14th, 2014, IIT Guwahati, Assam, India, 567 (1–6).

Chen, Z., & Tan, K. (2011). The perceived impact of JIT implementation on operations performance: Evidence from Chinese firms. *Journal of Advances in Management Research, 8*(2), 213–235.

Claycomb, C., Germain, R., & Dröge, C. (1999). Total system JIT outcomes: inventory, organization and financial effects. *International Journal of Physical Distribution and Logistics, 29*(10), 612–630.

Deif, A. M., & ElMaraghy, H. (2014). Cost performance dynamics in lean production leveling. *Journal of Manufacturing Systems, 33*(4), 613–623.

Green, K., Inman, R., Birou, L., & Whitten, D. (2014). Total JIT (T-JIT) and its impact on supply chain competency and organizational performance. *International Journal of Production Economics, 147*(Part A), 125–135.

Hobbs, D. P. (2011). *Applied lean business transformation: A complete project management approach.* Fort Lauderdale, FL: J. Ross Publishing.

Katayama, H., & Bennett, D. (1996). Lean production in a changing competitive world: A Japanese perspective. *International Journal of Operations & Production Management, 16*(2), 8–23.

Locher, D. A. (2008). *Value stream mapping for lean product development.* New York, NY: Productivity Press.

Losonci, D., & Demeter, K. (2013). Lean production and business performance: International empirical results. *Competitiveness Review, 23*(3), 218–233.

Monden, Y. (1993). *Toyota Production System: An integrated approach to just-in time* (2nd ed.). Norcross, GA: Industrial Engineering and Management Press, Institute of Industrial Engineers.

Myerson, P. (2012). *Lean supply chain and logistics management*. New York, NY: McGraw-Hill.

Narusawa, T., & Shook, J. (2009). *Kaizen express: Fundamentals for your lean journey*. São Paulo, Brasil: Lean Enterprise Institute.

Ohno, T. (1988). *Toyota production system: Beyond large scale production*. New York, NY: Productivity Press.

Phogat, S., & Gupta, A. (2017). Theoretical analysis of JIT elements for implementation in maintenance sector. *Uncertain Supply Chain Management, 5*(3), 187–200.

Shah, R., & Ward, P. T. (2007). Defining and developing measures of lean production. *Journal of Operations Management, 25*(4), 785–805.

Ward, A. C. (2007). *Lean product and process development*. Cambridge, MA: The Lean Enterprise Institute, Inc.

Wafa, M., & Yasin, M. (1998). A conceptual framework for effective implementation of JIT: An empirical investigation. *International Journal of Operations & Production Management, 18*(11), 1111–1124.

Wyrwicka, M., & Mrugalska, B. (2017). Mirages of Lean Manufacturing in Practice 7th International Conference on Engineering. *Project, and Production Management, Procedia Engineering, 182*, 780–785.

Womack, J. P., & Jones, D. T. (2003). *Lean thinking: Banish waste and create wealth in your corporation*. London: Simon & Schuster UK Ltd.

Womack, J. P., & Jones, D. T. (1996). *Lean thinking: Banish waste and create wealth in your corporation*. New York, NY: Simon and Schuster.

Womack, J. P., Jones, D. T., & Roos, D. (1990). *The machine that changed the world*. New York, NY: Macmillan Publishers.

Wong, Y. C., & Wang, K. (2011). Approaches and practices of lean manufacturing: The case of electrical and electronics companies. *African Journal of Business Management, 5*(6), 2164–2174.

Lean Manufacturing and Kaizen 2

2.1 Lean Manufacturing

Lean philosophy became popular through the book "*The Machine That Changed the World*", authored by Womack, Jones, and Ross (1990). While several definitions of lean manufacturing can be found in the literature (Dahlgaard & Dahlgaard-Park, 2006), all of them argue that lean manufacturing contribute to a higher quality of products and services, lower costs of production and the elimination of waste (Dennis, 2007; Hines, Holweg, & Rich, 2004). The United States (US) Environmental Protection Agency (EPA, 2003) defined the aims of lean as follows: "Develop the highest quality products, at the lowest cost, with the shortest lead time by systematically and continuously eliminating waste, while respecting people and the environment".

Lean has been analysed from different perspectives. From the philosophical perspective, lean is about eliminating waste from the production system and being able to produce products of high quality to satisfy the final customers (Ohno, 1988; Shingo, 1989). From the academic perspective, lean is considered to be a rule-driven system. From the professionals' perspective, lean is viewed as a congregation of tools and techniques aimed at eliminating waste (Shah & Ward, 2007).

Lean manufacturing is considered a strategic approach for companies in responding appropriately to a set of challenges, particularly market competition, increased pressure on inventory, increased service levels and reduced work in progress (WIP) (Laguna & Martí, 2001; Meade, Kumar, & Houshyar, 2006).

Although lean practices were created in the automobile industry, they are present in many companies from different sectors, such as forging companies (Sahoo, Singh, Shankar, & Tiwari, 2008), process industries (Abdulmalek & Rajgopal, 2006), clothing (Sukwadi, Wee, & Yang, 2013), hi-tech companies (Wang, 2008), aviation/aerospace (Ehret & Cooke, 2010; Kumar, Sharma, & Agarwal, 2015) and healthcare (Costa & Godinho Filho, 2016; Henrique, Rentes, Godinho Filho, & Esposto, 2016). Lean manufacturing has also been implemented together with other

J. L. Q. Pinto et al., *Just in Time Factory*, Management for Professionals,
https://doi.org/10.1007/978-3-319-77016-1_2

management approaches, such as the theory of constraints (Utiyama & Godinho Filho, 2013; Votto & Fernandes, 2014), sustainable manufacturing (Jabbour, Jabbour, Freitas, & Teixeira, 2013), cellular manufacturing (Bocanegra-Herrera & Orejuela-Cabrera, 2017) and the triple bottom line (Raj, Ma, Gam, & Banning, 2017).

The main concern of lean manufacturing is the reduction of wastes associated with machine setup time, inventory and other sources of waste through better control of production using kanban cards, developing partnerships with suppliers, optimizing the mix of production, investing in quality programmes and continuous improvement and the implementation of error-proof systems (Achanga, Shehab, Roy, & Nelder, 2006).

However, to ensure success, lean companies should implement a set of good practices such as improving continuous flow manufacturing, implementing customer-driven production, ensuring flexibility, waste elimination and zero defects, employing visual management, providing a safe and orderly working environment and ensuring customer-perceived quality (Womack & Jones, 2003). Beyond these, other factors are also identified as mandatory for success in lean manufacturing, such as: (1) high awareness of lean principles (Eswaramoorthi, Kathiresan, Prasad, & Mohanram, 2011), (2) recognition of the essence of lean and associated perceptions (Achanga et al., 2006), (3) understanding the organizational environment of the company in which lean manufacturing will be implemented (Agarwal, Shankar, & Tiwari, 2006), (4) identifying potential barriers to lean thinking and lean implementation (Gupta & Jain, 2013; Kumar & Mehta, 2013), (5) determining a suitable strategy for lean implementation (Karim & Arif-Uz-Zaman, 2013) and (6) developing strategic mechanisms to support lean transformation.

2.1.1 Foundations of Lean Manufacturing

During the last century, an important body of knowledge emerged on achieving superior quality (Gryna, Chua, & Defeo, 2007). W. Edwards Deming provided the foundation for the Japanese resurgence as an economic power after World War II (Besterfield, Michna, Besterfield, & Sacre, 2003). Deming made it his mission to teach optimal management principles for organizations to focus on quality. Deming's philosophy defends that improving quality would lead to decreased costs, less mistakes, less delays, better use of resources and improved productivity, thus enabling companies to obtain more market share, stay in business and add jobs (Summers, 2009).

Deming's philosophy was detailed within his 14 Points for Management, which he presented to the leaders of Japan in the 1950s and then to leaders in the US over later years (Besterfield et al., 2003).

Armand Feigenbaum is considered to be the originator of the total quality movement (Summers, 2009). Feigenbaum contributed to the quality movement by claiming that total quality control was necessary to achieve productivity, market penetration and competitive advantage (Besterfield et al., 2003). Summers (2009) wrote that Feigenbaum encouraged companies to eliminate waste, which drains profitability, by understanding the costs associated with failed quality levels. Feigenbaum emphasized the concepts of total quality control throughout all functions of the organization in order to ensure customer satisfaction and an economical cost of quality (Gryna et al., 2007).

Kaoru Ishikawa taught the Japanese problem-solving techniques and the use of high- quality tools. Ishikawa developed the "cause and effect diagram", used to find the root cause of a problem (Gryna et al., 2007). Ishikawa also developed "quality circles" whereby work groups were established to use high-quality tools for solving problems (Besterfield et al., 2003).

Phillip Crosby developed the notion of "doing it right the first time" in his 1979 book entitled "*Quality is Free*", which sold 1.5 million copies. Crosby developed the "four absolutes of quality management", which set expectations to be met through a continuous improvement process (Besterfield et al., 2003). Crosby defined four absolutes of quality management in order to manage quality: prevention of defects, zero defects or making products right the first time and reducing or eliminating the costs associated with poor quality (Summers, 2009).

Genichi Taguchi developed a concept called the "loss function", which combines cost, target and variation into one metric (Besterfield et al., 2003). Taguchi's concept described product variation from a target dimension resulting in a total loss to society. According to Taguchi's loss function, any deviation from a target, even if within specification, would result in reduced quality levels and a loss to society as a whole (Summers, 2009).

The concept of continuous improvement originated with US companies such as the National Cash Register and Lincoln Electric Company, dating as far back as 1894 (Westcott, 2006). Japanese companies such as Toshiba, Matsushita Electric and the Toyota Motor Company began the development of continuous quality improvement programmes in the early 1950s (Westcott, 2006).

Manufacturing progressed from craftsman production in the early 1900s to mass production based on the Fred Winslow Taylor system of separating planning from production (Dennis, 2007). Although the Taylor system had a reputation for mindless and dehumanizing work, it did uncover lean-related innovations, such as standardized work, shorter cycle times, time and motion studies and measurement and analysis (Dennis, 2007).

A gap developed between management and the production worker and accounting practices encouraged wasteful manufacturing activities, such as building inventory rather than building in line with customer demand. The growing dysfunction resulted in worker alienation, poor quality, excessive machinery and engineering functionality issues (Dennis, 2007). The birth of lean production dates back to the 1950s after a Japanese engineer named Eiji Toyoda visited Ford's Rouge plant in Detroit. Once he returned to Japan, Toyoda and Taiichi Ohno concluded that mass production would not work in Japan. Ohno went to work on developing lean principles and whereas years earlier Taylor had separated planning and production, Ohno brought them back together again (Dennis, 2007).

Taiichi Ohno (1988) defined the topic of lean in his seminal publication entitled "*The Toyota Production System*". Liker (2004) described the TPS as the basis for the lean production movement, which has dominated manufacturing around the world.

2.1.2 Main Lean Principles and Wastes

As stated above, there are several elements considered essential to the success of lean manufacturing. Among these are the reduction of machine setup times, inventory and waste reduction, responsiveness of manufacturing, the development of partnerships with suppliers, the development of quality and improvement programmes and the implementation of foolproof systems.

To reduce setup times, lean manufacturing suggests the single-minute exchange of die (SMED) technique to guarantee quick changeover and reduce inventories and total lead times (Shingo, 1989). In terms of inventory and waste reduction lean manufacturing suggests the implementation of a non-wasteful and responsible manufacturing system. This is because clients are now willing to pay only for value-added activities that contribute something to the products and services (Shook & Rother, 2003). Manufacturing should also be responsive to allow the implementation of pull production systems, avoiding waste resulting from the production of a final stock of items (Womack & Jones, 2006). To this end, the lean tool Kanban is suggested as a means of pulling the processes, from the customer to the supply of raw material.

Partnerships with suppliers represents another determinant factor for the success of lean manufacturing, allowing closer relationships between companies and their suppliers in terms of sharing developments, dividing responsibilities and implementing accurate delivery and high-quality standards through JIT production (Ohno, 1988). Another factor that contributes to the success of lean manufacturing is the development of quality and improvement programmes (kaizen) between supply chain partners. Using statistical techniques, these programmes increase the empowerment of workers and thus contribute to better performance of processes (Myerson, 2012; Timans & Antony, 2012).

Finally, the implementation of foolproof systems (called poka-yoke in Japanese) contributes to eliminating possible failures and ensuring the production of highly reliable products (Ohno, 1988).

To succeed, lean manufacturing should be implemented respecting the following set of principles (Womack and Jones (2003): (1) specify value based on the perception of the end customer and by product family; (2) identify all the steps in the process value stream for each product family, eliminating those steps that do not add value; (3) ensure the value-creating steps occur in a tight sequence to secure the smooth flow of the product towards the customer; (4) as flow is established, let customers pull value from the next upstream activity; (5) once value is specified, identify value streams, eliminate wasteful steps and establish flow and pull.

Table 2.1 Toyota's management principles

	Sections		Principles
1	Long-term philosophy	1	Base your management decision on a long-term philosophy, even at the expense of short-term financial goals
2	The right process will produce the right results	2	Continuous process flow
		3	Use pull systems to avoid overproduction
		4	Level out the work load (Heijunka)
		5	First time quality right the first time (Jidoka)
		6	Standardized tasks
		7	Use visual controls
		8	Reliable equipment
3	Add value to the organization by developing your people and partners	9	Grow leaders from within
		10	Develop exceptional people
		11	Respect your partners and help them improve
4	Continuously solving root problems drive organizational learning	12	Go and see for yourself (Genchi Gebbutsu)
		13	Make decision slowly considering all options
		14	Become learning organizations through reflection and Kaizen

Source: Liker (2004, pp. 37–41)

Liker (2004) identified 14 management principles in his book entitled "*The Toyota Way*". These management principles are illustrated in Table 2.1 by section and principle.

2.1.2.1 Lean Waste

Lean waste is designated in Japanese as *muda* and is considered by Russell and Taylor (2000, p. 234) to constitute "anything other than the minimum amount of equipment, materials, parts, space and time which are essential to add value to the product". According to Ohno (1988) there are seven types of waste: overproduction, excess inventory, waiting, unnecessary motion, excess transportation, rejections/reworking and over-processing. Environmental waste has recently been considered one of the "new" wastes associated directly with wasted materials (Bicheno, 2000). The elimination of these wastes seems simple and straightforward, but their identification is often difficult in most organizations. Table 2.2 provides insights into the definition of the different types of waste considered in lean philosophy.

Lean manufacturing represents an important contribution to the better performance of industrial companies by eliminating waste through reducing supplier, customer and internal variability (Angelis & Fernandes, 2012), reducing production costs and accelerating product lead times (Chiarini, 2012, 2013).

2.1.3 Benefits of the Implementation of Lean Manufacturing

Silva, Santos, and Castro (2012) argue that a sound manufacturing strategy contributes for making a company more competitive by providing a competitive

Table 2.2 Types of waste

Type of waste	Description
Product defects	Products that are not to specifications, resulting in reworking, scrapping, check-ups, flawed materials or replacement production.
Waiting time	Non-productive time leading to postponements allied with stock-outs, equipment downtime, machine breakdown, capacity bottlenecks and lot-processing delays.
Processing waste	Process phases not essential to attain the final product.
Over-production	A level of production higher than demand and before it is needed.
Motion waste	The unnecessary motion of people on the shop floor, i.e. not supporting the manufacture of the product.
Inventory	Non-optimal storage of final products, intermediates or raw materials; these are expensive so they should be maintained at the lowest level possible.
Transportation	Excess of product transportation not counting as value to the client.

Source: Zakaria, Mohamed, Ab Rahid, and Mohd (2017)

priority that is perceived by its customers. Alves (2001) emphasizes the importance of the strategic vision of the production department in choosing the right management model for the production process in order to support the company in developing and maintaining a competitive advantage. According to Singh and Mahmood (2014), the best management model for ensuring high-performance manufacturing processes should be one based on high-quality products, processing speed, cost, flexibility and reliability. In this context, lean manufacturing, suggesting continuous improvement' philosophy and waste reduction, contributes for increasing the competitiveness of companies in a specified industry (Lucato, Vieira Jr., Vanalle, & Salles, 2012).

The multiple definitions of lean manufacturing, besides clarifying the concept, also identify the main objectives/advantages associated with its implementation, such as the following: inventory minimization, the implementation of managerial practices in production, increasing value, minimizing waste. Besides these advantages, Womack and Jones (2003) highlight productivity and customer satisfaction. In addition, Karlsson and Åhlström (1996) consider as the main benefits to be improved quality, speed and cost reduction and Moyano-Fuentes and Sacristán-Díaz (2012) the increased efficiency of the value creation process. Lean manufacturing has been widely used by companies in all kinds of industrial sectors around the world because of the benefits they can attain with its implementation.

A positive relationship between the implementation of lean manufacturing and business performance has been highlighted in many studies. For example, in the work of Losonci and Demeter (2013), it is evident that the implementation of lean manufacturing exerts a significant influence on a set of performance indicators, such as inventory turnover, quality, production time, labour productivity, space utilization, production volume flexibility, product mix and costs. Also, Melton (2005) and Rose, Deros, Ab Rahman, and Nordin (2011) demonstrated the positive impact of lean manufacturing on performance, highlighting the following associated benefits: financial savings, less reworking, reduced lead times, increased process understanding, less process waste and reduced inventory.

The impact of lean manufacturing on inventory and waste is widely recognized in many studies concerning lean manufacturing and performance based on the argument that the customers agree on to pay only for added-value activities associated to the product or service. Being so, the company should implement a non-wasteful and responsive manufacturing system (Hofer, Eroglu, & Rossiter, 1995). Moreover, the reduction in costs is justified by the elimination of non-added activities and by linking and balancing work stages.

Many surveys have also been undertaken to analyse the impact of lean manufacturing on performance. For example, Chavez et al., (2015) investigated the impact of lean manufacturing on operational and organizational performance, concluding that there is a positive association between them, mainly supported by supply chain relationships. Another interesting study was that of Azadegan, Patel, Zangoueinezhad, and Linderman (2013), investigating the effect of environmental uncertainty on the relationship between lean practices and performance. They concluded that environmental complexity moderates in a positive way the effects of lean practices on performance, but that environmental dynamism reduces the benefits of lean operations.

Focusing only on the relationship between lean manufacturing and inventory turnover and attending to production systems, order types and product types, Demeter and Matyusz (2011) concluded that companies implementing lean manufacturing attain better inventory turnover for each type of inventory than traditional companies. Also, King and Lenox (2001) studied the impact of lean manufacturing on the environment in terms of pollution and waste costs and concluded that it has a positive impact.

Johana et al. (2014) focused on a specific lean tool, 5S (sort, set in order, shine, standardize and sustain), to study its impact on the workplace environment and concluded that it provides a more comfortable, safe and organized workplace. Uma and Kannan (2010) studied the implementation of the 5S lean tool in human resource management (HRM) and found that the employees of organizations fail to change while customer demand and market expectations change rapidly. In 2010, a study was conducted examining the implementation of 5S practices in manufacturing companies (Rahman, Khamis, Zain, Deros, & Mahmood, 2010). An 5S checklist was developed for each company for the auditing process. Both the companies associated with the research performed excellent 5S practices. This study demonstrated that 5S contributes for improving housekeeping and enhancing health and safety standards in the workplace, ensuring also a better environmental performance. The study also highlights the effort and participation of top management as a mandatory factor for a successful implementation of 5S (Rahman et al., 2010).

Lean manufacturing contributes for increasing benefits associated to inventory, manufacturing efficiency, improved quality, increased flexibility and improved customer satisfaction (Worley & Doolan, 2006). The heart of lean and the TPS is the elimination of waste (Liker, 2004). Careful inspection of any production area reveals waste and room for improvement (Ohno, 1988, p. 55). As argued by Shaw and Ward (2003, p. 129), "the core thrust of lean production is that lean practices can work synergistically to create a streamlined, high quality system that produces finished products at the pace of customer demand with little or no waste".

In a study conducted by Keyser and Sawhney (2013), it was found that a company named the Knoxville Corrugated Box Company recognized success attained by changing from a non-lean enterprise to a lean manufacturing firm. This included a 33% reduction in inventory levels, an improvement in on-time delivery performance from 75% to 96% and a 50% reduction in customer complaints. Kumar and Abuthakeer (2012) found that some organizations could reduce their machine set-up times from 25% to 85% by using the SMED lean technique. This leads to greater production flexibility and machine utilization. Also, Zayco, Broughman, and Hancock (1997) concluded in their work that lean manufacturing contributes to reduce in half the human effort, decrease the manufacturing space, improve tool investment and product development time, as well as a substancial improvement in quality (Wu, 2003). A study of 200 US manufacturers demonstrated improved inventory and financial and market performance due to lean methods (Wu, 2003). The use of lean initiatives in a small manufacturing firm resulted in important and strategic improvements in cycle time, change over time and in lead time (Grewal, 2008).

The main benefits associated with the implementation of lean manufacturing, as focused on above, are summarized in Table 2.3.

There are a lot of researches about the effects of lean manufacturing on performance, however they present some inconsistencies in their results with some of them sowing a positive effect of lean on performance (Shah & Ward, 2003), but other researches argue that the influence of lean practices on performance depends on some external factors (Ketokivi & Schroeder, 2004; Swink, Narasinhan, & Kim, 2005). This could be justified by different contexts and environmental factors, such as cultural factors, company characteristics, inter alia, that may affect the way lean manufacturing influences organizational performance. Considering cultural factors, lean manufacturing represents a set of techniques, methods and practices that are conceptually different from traditional production systems. Its principal concern is to identify and eliminate waste representing a new and revolutionary way of thinking about the manufacturing process. For that, a cultural change is mandatory in order to achieve the discipline necessary to apply these new concepts and principles, thus guaranteeing the successful implementation of the lean manufacturing system (Alves, 2010).

Lean manufacturing is a strategic alternative to the traditional mass production system widely adopted by various industries (Ramesh & Kodali, 2012) to attain productivity gains. However, its implementation commonly requires physical changes associated with layout changes, the implementation of lean tools and the development of ways of enabling smooth production, inventory reduction and standardized methods.

Lead-time performance represents a major competitive priority for manufacturers from different industries (Treville, Shapiro, & Hameri, 2004). Two approaches have

Table 2.3 Benefits of lean manufacturing

Benefits	Authors
Financial savings	Melton (2005), Rose et al. (2011), Wu (2003)
Less reworking	Melton (2005), Rose et al. (2011)
Reduced lead time	Losonci and Demeter (2013), Melton (2005), Rose et al. (2011), Wu (2003)
Improved housekeeping and health and safety standards in the workplace	Rahman et al. (2010), Zayco et al. (1997)
Improved production flexibility	Kumar and Abuthakeer (2012)
Increased quality	Shaw and Ward (2003), Worley and Doolan (2006), Zayco et al. (1997)
Increased process understanding	Melton (2005), Rose et al. (2011)
Less process waste	Melton (2005), Rose et al. (2011)
Reduced inventory	Keyser and Sawhney (2013), Melton (2005), Rose et al. (2011), Worley and Doolan (2006), Wu (2003)
Improved responsiveness of companies	Hofer et al. (1995), Keyser and Sawhney (2013), Worley and Doolan (2006)
Cost reduction	Elmoselhy (2013)
Reduced cycle time	Wu (2003)
Reduced product development time	Zayco et al. (1997)
Positive impact on business performance	Losonci and Demeter (2013), Moori, Pescarmona, and Kimura (2013)
Waste and pollution reduction	King and Lenox (2001)
Better manufacturing efficiency	Worley and Doolan (2006)
Improved machine utilization	Kumar and Abuthakeer (2012)
Improvement of operational performance and organizational performance	Azadegan et al. (2013), Chavez et al. (2015), Shah and Ward (2003)
Reduced machine set-up times	Kumar and Abuthakeer (2012)
Better inventory turnover	Demeter and Matyusz (2011), Losonci and Demeter (2013)
Improvement of productivity	Losonci and Demeter (2013)
Improved customer satisfaction	Keyser and Sawhney (2013), Worley and Doolan (2006)

been identified as contributing to reducing lead times in manufacturing: The integration of information technology (IT) within and between firms in the supply chain and the improvement of process through lean/JIT manufacturing practices have contributed to reducing lead times. Some studies in time-based manufacturing (Nahm, Vonderembse, & Koufteros, 2003) have identified certain practices within lean manufacturing that have a positive impact on lead time and firm performance, contributing to making organizations more customer-oriented (Fisher, Raman, & McClelland, 2000) and consequently more competitive (Rondeau, Vonderembse, & Ragu-Nathan, 2000).

2.2 Kaizen

2.2.1 Fundamentals of Kaizen

In order to improve their competitiveness one of the strategies that many companies adopt is to employ continuous improvement or kaizen in their organizations (Teece, 2007). The meaning of the word kaizen is change for the better (Coimbra, 2009). Kaizen also means continuing improvement in a cooperative way, by incorporating everyone in an organization, such as managers and workers, without spending much money (Imai, 1997). This means that everybody in the organization should constantly be looking for better ways to execute their jobs by identifying and eliminating muda (or non-value-adding activities), so as to streamline the work processes (Imai, 1997). Thus, continuous improvement should be monitored and implemented daily. Bhuiyan and Baghel (2005) define kaizen as "a culture of sustained improvement targeting the elimination of waste in all systems and processes of an organization". Kaizen is supported through several problem-solving tools focused on the identification and resolution of work-based problems. The aim is to attain new "benchmarks" for every single problem that is solved. Moreover, to consolidate the new "benchmarks", they should be standardized (Terziovskia & Sohalb, 2000).

Kaizen is a philosophy originating from Japan that encourages small improvements obtained through continuing effort. Since World War II, this methodology has been implemented successfully by Japanese industry (Imai, 1986). Moreover, Toyota has developed kaizen over the last years in its Toyota way of management, based on which continuous improvement is a daily occurrence (Coimbra, 2009). Long-term improvement is obtained by motivating the employees to work towards gradually higher work standards. At Toyota, the improvement process takes place every day within every process (activity) and at every level of the enterprise. Moreover, this improvement does not stop even if the numbers have already been met. Of course, from day to day, improvement may involve small steps (Rother, 2010).

Kaizen enables people to be flexible in relation to the changes occurring, thus incorporating such changes in their daily routine actions (De De Lange-Ros & Boer, 2001). Kaizen symbolizes the mobilization of the workforce, acting as the main channel used by employees to contribute to their company's development (Brunet & New, 2003).

Imai (1986) introduced the kaizen umbrella concept, under which several methods and approaches are incorporated, including customer orientation, total quality management (TQM), robotics, quality control circles (QCCs), suggestion systems, automation, discipline in the workplace, total productive maintenance (TPM), kanban, quality improvement, JIT, zero defects, productivity improvement and new product development, the three Mus, genchi genbutsu and brainstorming (Imai, 1986; Kiran, 2017). In a study of Japanese manufacturing plants, Cheser (1998) concluded that an increase in motivation and a positive change in employees' attitudes occurred after the introduction of kaizen.

Kaizen should encourage employees to improve the workplace and contribute to greater self-reliance and self-control. The basic objective of kaizen is to improve the following three parameters: quality, cost and delivery time. The important elements of the kaizen concept are ensuring standardization and creating systems for maintaining the level of change. In practice, the definition of standards means their continuous enhancement as a result of continuous kaizen activities. However, it is necessary to be aware that the application of this tool does not exclude the introduction of radical improvements that can be supported and fixed by using kaizen.

With the aim of developing a conclusive and insightful working definition of kaizen for practitioners and academics in the West, Macpherson, Lockhart, Kavan, and Iaquinto (2015) explored how Japanese workers acknowledge, exercise, identify and diffuse kaizen in a sustainable manner. The authors concluded that two demonstrations of the kaizen philosophy emerge among the Japanese people: (1) in daily life, in which kaizen points to improvement and ingenuity; (2) in industrial sets, in which it refers to management's commitment to the pursuit of business excellence through the interplay of the enterprise-side pursuit of profit and competition and employee-side skills, creativity, confidence and pride. Also, besides human traits, kaizen also involves the means necessary to operate—tools and methods—and to generate and implement improvement. The integration of these two elements—the company/employee side and tools and methods—results in an energy that permeates the company and creates a shared state of mind among employees to achieve proactive change and innovation. More than being simply about continuous improvement, as often referred in the Western literature, kaizen is the means and the result of the demands of management and the management of human and non-human resources in the organization's pursuit of business excellence (Macpherson et al., 2015).

The main characteristics of kaizen emphasized in the literature are that: (1) kaizen is continuous, which is used to denote its embedded nature and also its place in an endless journey towards quality and efficiency; (2) kaizen is incremental, in contrast to major reorganizations typically initiated by management or to technological innovation (e.g. the installation of new technology or machinery); (3) kaizen is participatory, demanding the involvement and intelligence of the workforce, generating intrinsic psychological benefits and quality of work and life for employees (Brunet & New, 2003). Nonetheless, continuous improvement can take place through evolutionary improvement, meaning that improvements are incremental, or though radical changes that result from an innovative idea or new technology. Frequently, major improvements occur over time as a result of abundant incremental improvements. In either case, improvement is supported by a set of tools and techniques capable of identifying the sources of problems, waste and variation, as well as seeking ways to eliminate or minimize these (Bhuiyan & Baghel, 2005).

The establishment and maintenance of continuous improvement demands a major, long-term effort (Caffyn, 1999) because kaizen requires breaking the traditional mind-set and encouraging employees to adapt to a new mind-set. To support the desired behaviours, typically the plan–do–check–act (PDCA) problem-solving cycle is used (Deming, 1986; Shewhart, 1939). Indeed, this simple but very systematic format of a "wheel" of never-ending improvements is the primary symbol for kaizen, in which the PDCA cycle acts as a standardization of the improvement process (Berger, 1997).

According to Imai (1986), continuous improvement should be practised through three types of kaizen: management-oriented kaizen, group-oriented kaizen and individual-oriented kaizen. The first type, management-oriented kaizen, is the most important since it entails the company strategy and the most important logistic aspects and involves managers and professionals. Group-oriented kaizen is represented by quality circles, in which employees are organized in teams or circles with the goal of finding and solving problems that occur during their day-to-day work, using statistical tools. The activities of group-oriented kaizen are typically confined to problems arising within the same workshop. Finally, individual-oriented kaizen involves everybody in the company and is manifested through a suggestions system, in which the employees can provide a solution or a recommendation to a given problem encountered. The employees come up with the ideas and their implementation is left to the specialists. This type of kaizen generally occurs within each working area.

In Burton and Boeder's (2003) work, this hierarchy is complemented with the introduction of the concept of kaizen events. A distinction is made between project kaizen events and process kaizen events. Project kaizen events are defined by the authors as events that focus on value stream improvement activities, enhancing value across multiple functions within a value stream. Typically, these project kaizen events entail cross-functional teams and comprise project-based continuous improvement activities. These events have a strategic importance for the organization, suppliers and customers, spanning a duration of many weeks to complete the kaizen initiative. In contrast, process kaizen events are focused on the elimination of waste in activities that are of limited definition. The main emphasis is on waste elimination, workplace effectiveness, or work standardization. These kaizen events are initiated as employee projects for the purpose of rapid hit process improvements. The terms used to describe these types of events are either kaizen blitz or kaizen super blitz. Kaizen blitz events correspond to short duration continuous improvement projects that are conducted over one to five days, in which a cross-functional multilevel team works intensely to develop, test and refine solutions to problems promptly and leave a new process in place. Kaizen super blitz events should be completed in one to eight hours and are usually prompted by the occurrence of a defect, failure, or safety issue. In this type of kaizen event, a cross-functional team works intensely to determine the root cause of the problem and implement short-term corrective actions. In Fig. 2.1 the three types of kaizen events are distinguished.

Caffyn (1999) propose a continuous improvement capacity model that describes continuous improvement as a set of key behaviours or behavioural routines which seem to be critical for long-term success with continuous improvement and development over time. Moreover, the author defines continuous improvement capability as "the ability of an organization to gain strategic advantage by extending involvement in innovation to a significant proportion of its members". The model is

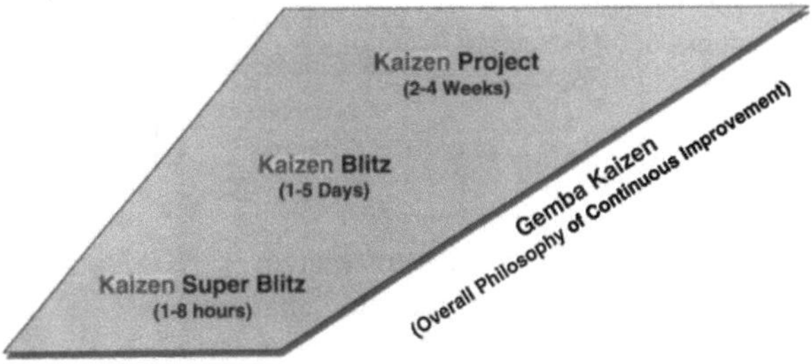

Fig. 2.1 Main obstacles to lean implementation. Source: Wyrwicka and Mrugalska (2017)

comprised by a set of 10 generic continuous improvement behaviours that are seen as fundamental in organizations of all types and sizes:

1. Employees demonstrate awareness and understanding of the organization's aims and objectives.
2. Individual groups consider the organization's strategic goals and objectives to prioritize their improvement activity.
3. The enabling mechanisms (e.g. training, teamwork) used to encourage involvement in continuous improvement are monitored and developed.
4. Ongoing assessment ensures that the organization's structure, systems and procedures and the approach and mechanisms used to develop continuous improvement constantly reinforce and support each other.
5. Managers at all levels display active commitment to and leadership of continuous improvement.
6. Throughout the organization, people engage proactively in incremental improvement.
7. There is effective working across internal and external boundaries at all levels.
8. People learn from their own and others' experiences, both positive and negative.
9. The learning of individuals and groups is captured and deployed.
10. People are guided by a shared set of cultural values underpinning continuous improvement as they go about their everyday work.

In Kiran's (2017) work a set of 10 requirements for continuous improvement success is identified:

1. Accepting that the problem lies in the inadequacy of the present level of performance of the product or service.
2. Right attitude to solving the problem.
3. Proper organization for solving the problem.
4. Adequate knowledge and practice in using problem-solving tools and techniques.
5. Structured method of problem solving.
6. Problem definition and analysis to be based on hard facts.
7. Solutions to the cause more than the symptoms of the problem.
8. Implementing and continuous monitoring until consistent results are obtained.

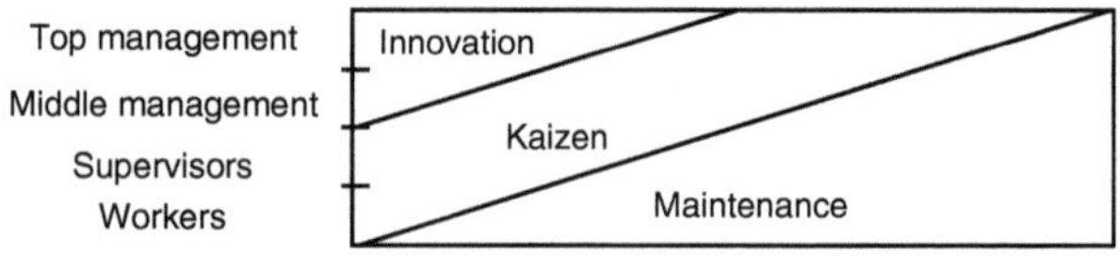

Fig. 2.2 Japanese perceptions of job functions. Source: Imai (1986)

9. Overcoming resistance to change.
10. Control system for reversible changes.

2.2.2 Kaizen and Innovation

In his seminal work on kaizen, Imai (1986) explains how job functions are seen in Japan along the levels of the organizational hierarchy (Fig. 2.2). According to the author, the job functions involve three primary components: maintenance, kaizen and innovation. Maintenance is connected to activities directed towards sustaining current technological, managerial and operating standards through training and discipline. This implies that management first establishes policies, rules, directives and procedures for all major operations and then see to it that everyone follows the established standard operating procedures. In contrast, improvement refers to improving standards and can be separated in kaizen and innovation. While kaizen is associated with continuous incremental improvements, innovation involves drastic improvements as a result of investments in new technology and/or equipment. Innovation can broadly be defined as new ideas that are actually applied in fundamentally different ways to generate new and/or additional value (Lee & Trimi, 2016).

Innovation has been classified in several different ways in the literature. However, many studies have suggested the following four broad classifications: incremental (exploitative), radical (explorative), ambidextrous and disruptive innovation (Lee & Trimi, 2016). Moreover Lee and Trimi (2016) point out that incremental innovation involves continuous improvement of what is already known and therefore kaizen can be seen as incremental innovation.

There are some significant differences between kaizen and innovation. Kaizen attains high performance levels at no or marginal costs, as opposed to innovation, which needs heavy investment. Also, kaizen emphasizes continuous improvement compared to innovation, which is a one-time improvement (Kiran, 2017). However, kaizen implies continuous effort and commitment at all levels of management. Thus kaizen requires a substantial commitment of time and effort on the part of management. Investing in kaizen means investing in people (Terziovskia & Sohalb, 2000). Nevertheless, the existence of kaizen does not diminish the importance and necessity of innovation. Indeed, innovation should be complemented by continuous improvement to ensure that the benefits of innovations increase over time rather than decreasing their utility due to a constantly changing environment (Chary, 2009). The Productivity Press Development Team (1992) considers that kaizen cannot

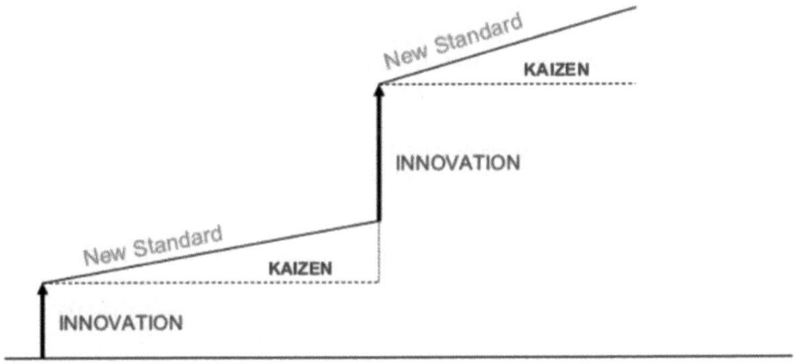

Fig. 2.3 Innovation plus kaizen. Source: Imai (1986)

thrive without innovation. Also, there is nothing more wasteful than trying to activate a kaizen system in a company that is not engaged in innovation.

The difference between the two opposing concepts may be likened to that of a staircase and a slope (Imai, 1986), as depicted in Fig. 2.3.

As Fig. 2.3 shows, whenever an innovation is achieved and is followed by a series of kaizen efforts to maintain and improve it, there is added value to the innovation process. Moreover, whenever a new innovation is generated, it has a starting point that is superior to the previous innovation. Thus, kaizen will target improvement until the next innovation to prevent a decline in the new standard. Indeed, whenever an innovation is introduced, without kaizen it will gradually deteriorate and when another innovation is introduced, the starting point will be lower than the point achieved in the previous innovation.

References

Abdulmalek, F. A., & Rajgopal, J. (2006). Analyzing the benefits of lean manufacturing and value stream mapping via simulation: A process sector case study. *International Journal of Production Economics, 107*, 223–236.

Achanga, P., Shehab, E., Roy, R., & Nelder, G. (2006). Critical success factors for lean implementation within SMEs. *Journal of Manufacturing Technology Management, 17*, 460–471.

Agarwal, A., Shankar, R., & Tiwari, M. K. (2006). Modeling the metrics of lean, agile and leagile supply chain: An ANP-based approach. *European Journal of Operational Research, 173*(1), 211–225.

Alves, J. M. (2001). *Proposta de um modelo híbrido de gestão da produção: aplicação na indústria aeronáutica* [*Proposal for a hybrid model of production management: Application in the aircraft industry*]. Campinas: Tese (Doutorado em Engenharia Mecânica), Faculdade de Engenharia Mecânica, Universidade Estadual de Campinas.

Alves, J. R. X. (2010). *Aplicação de um modelo de gestão da manufatura enxuta com foco na transformação cultural: aplicação em uma empresa da indústria automotiva* [Application of a management model of lean manufacturing with focus on cultural transformation: Application in a company of the automotive industry]. São José dos Campos: Tese (Mestrado em Engenharia Aeronáutica e Mecânica), Instituto Tecnológico de Aeronáutica.

Angelis, J., & Fernandes, B. (2012). Innovative lean: Work practices and product and process improvements. *International Journal of Lean Six Sigma, 3*(1), 74–84.
Azadegan, A., Patel, P. C., Zangoueinezhad, A., & Linderman, K. (2013). The effect of environmental complexity and environmental dynamism on lean practices. *Journal of Operations Management, 31*, 193–212.
Berger, A. (1997). Continuous improvement and *kaizen* : Standardization and organizational designs. *Integrated Manufacturing Systems, 8*(2), 110–117. https://doi.org/10.1108/09576069710165792
Besterfield, D. H., Michna, C. B., Besterfield, G. H., & Sacre, M. B. (2003). *Total quality management*. Upper Saddle River, NJ: Prentice Hall.
Bhuiyan, N., & Baghel, A. (2005). An overview of continuous improvement: From the past to the present. *Management Decision, 43*(5), 761–771.
Bicheno, J. (2000). *The new lean toolbox: Towards fast flexible flow*. Buckingham: PICSIE Books.
Bocanegra-Herrera, C. C., & Orejuela-Cabrera, J. P. (2017). Cellular manufacturing system selection with multi-lean criteria, optimization and simulation. *Ingenieria y Universidad, 21*(1), 7–26.
Brunet, A., & New, S. (2003). Kaizen in Japan: An empirical study. *International Journal of Operations and Production Management, 23*(12), 1426–1446. https://doi.org/10.1108/01443570310506704
Burton, T., & Boeder, S. (2003). *The lean extended enterprise: Moving beyond the four walls to value stream excellence*. Boca Raton, FL: J. Ross Publishing.
Caffyn, S. (1999). Development of a continuous improvement self-assessment tool. *International Journal of Operations and Production Management, 19*(11), 1138–1153.
Chary, S. (2009). *Production and operations management* (4th ed.). New Delhi: Tata McGraw-Hill.
Chavez, R., Yu, W., Jacobs, M., Fynes, B., Wiengarten, F., & Lecuna, A. (2015). Internal lean practices and performance: The role of technological turbulence. *International Journal of Production Economics, 160*, 157–171.
Cheser, R. (1998). The effect of Japanese kaizen on employee motivation in US manufacturing. *International Journal Organization, 6*(3), 197–217.
Chiarini, A. (2012). Lean production: Mistakes and limitations of accounting systems inside the SME sector. *Journal of Manufacturing Technology Management, 23*(5), 681–700.
Chiarini, A. (2013). Building a six sigma model for the Italian public healthcare sector using grounded theory. *International Journal of Services and Operations Management, 14*(4), 491–508.
Coimbra, E. (2009). *Total flow management: Achieving excellence with kaizen and lean supply chains*. Switzerland: Kaizen Institute.
Costa, L. B. M., & Godinho Filho, M. (2016). Lean healthcare: Review, classification and analysis of literature. *Production Planning and Control, 27*(10), 823–836.
Dahlgaard, J. J., & Dahlgaard-Park, D. M. (2006). Lean production, six sigma quality, TQM and company culture. *TQM Magazine, 18*(3), 263–281.
De Lange-Ros, E., & Boer, H. (2001). Theory and practice of continuous improvement in shop-floor teams. *International Journal of Technology Management, 22*(4), 344–358.
Demeter, K., & Matyusz, Z. (2011). The impact of lean practices on inventory turnover. *International Journal of Production Economics, 133*, 154–163.
Deming, W. E. (1986). *Out of the crisis*. Cambridge, MA: Massachusetts Institute of Technology Center for Advanced Engineering Study.
Dennis, P. (2007). *Lean production simplified*. New York, NY: Productivity Press.
Ehret, O., & Cooke, P. (2010). Conceptualising aerospace outsourcing: Airbus UK and the lean supply approach. *International Journal of Technology Management, 50*(3/4), 300–317.
Elmoselhy, S. A. M. (2013). Hybrid lean-agile manufacturing system technical facet, in automotive sector. *Journal of Manufacturing Systems, 32*(2013), 598–619.
EPA. (2003). *Lean manufacturing and the environment: Research on advanced manufacturing systems and the environment and recommendations for leveraging better environmental performance*. http://www.epa.gov/lean/environment/pdf/leanreport.pdf
Eswaramoorthi, M., Kathiresan, G. R., Prasad, P. S. S., & Mohanram, P. V. (2011). A survey on lean practices in Indian machine tool industries. *International Journal of Advanced Manufacturing Technology, 52*, 1091–1101.

Fisher, M., Raman, A., & McClelland, A. S. (2000). Rocket science retailing is almost here—Are you ready? *Harvard Business Review, 78*(4), 115–124.

Grewal, C. S. (2008). An initiative to implement lean manufacturing using value stream mapping in a small company. *International Journal of Manufacturing Technology and Management, 15*(3–4), 404–417.

Gryna, F. M., Chua, R. C. H., & Defeo, J. A. (2007). *Juran's quality planning and analysis for enterprise quality*. New York, NY: McGraw Hill.

Gupta, S., & Jain, S. K. (2013). A literature review of lean manufacturing. *International Journal of Management Science and Engineering Management, 8*, 241–249.

Henrique, D. B., Rentes, A. F., Godinho Filho, M., & Esposto, K. F. (2016). A new value stream mapping approach for healthcare environments. *Production Planning and Control, 27*(1), 24–48.

Hines, R., Holweg, M., & Rich, N. (2004). Learning to evolve: A review of contemporary lean thinking. *International Journal of Operations and Product Management, 24*(10), 994–1011.

Hofer, C., Eroglu, C., & Rossiter, H. A. (1995). The effect of lean production on financial performance: The mediating role of inventory leanness. *International Journal of Production Economics, 138*(2), 242–253.

Imai, M. (1986). *Kaizen: The key to Japan's competitive success*. New York: McGraw-Hill.

Imai, M. (1997). *Gemba kaizen: A commonsense, low-cost approach to management*. New York: McGraw-Hill.

Jabbour, A. B. L. S., Jabbour, C. J. C., Freitas, W. R. S., & Teixeira, A. A. (2013). Lean and green? Evidências empíricas do setor automotivo Brasileiro. *Revista Gestão E Produção, 20*(3), 653–665.

Johana, Y., Hardi, N. M., Abdullah, L., Jumadi, N., Mohammad, W. N. S. W., & Taharuddin, N. S. (2014). *The sustainability of QE/5S implementation in an administration office of a higher education institution*. Retrieved from https://www.researchgate.net/publication/278244980

Karim, A., & Arif-Uz-Zaman, K. (2013). A methodology for effective implementation of lean strategies and its performance evaluation in manufacturing organizations. *Business Process Management Journal, 19*(1), 169–196.

Karlsson, C., & Åhlström, P. (1996). Assessing changes towards lean production. *International Journal of Operations & Production Management, 16*, 24–41.

Ketokivi, M., & Schroeder, R. G. (2004). Manufacturing practices, strategic fit and performance: A routine based view. *International Journal of Operations and Production Management, 24*(2), 171–191.

Keyser, R. S., & Sawhney, R. S. (2013). Reliability in lean systems. *International Journal of Quality and Reliability Management, 30*(3), 223–238.

King, A. A., & Lenox, M. J. (2001). Lean and green? An empirical examination of the relationship between lean production and environmental performance. *Production and Operations Management, 10*(3), 244–256.

Kiran, D. R. (2017). *Total quality management: Key concepts and case studies*. Oxford, UK: Butterworth Heinemann.

Kumar, B. S., & Abuthakeer, S. S. (2012). Implementation of lean tools and techniques in an automotive industry. *Journal of Applied Sciences, 12*(10), 1032–1037.

Kumar, G. A., & Mehta, R. K. (2013). Lean manufacturing practices: Emerging issues. *International Journal of Innovative Research and Development, 2*, 69–78.

Kumar, R. B. R., Sharma, M. K., & Agarwal, A. (2015). An experimental investigation of lean management in aviation – Avoiding unforced errors for better supply chain. *Journal of Manufacturing Technology Management, 26*(2), 231–260.

Laguna, M., & Martí, R. (2001). Neural network prediction in a system for optimizing simulations. *IIE Transactions, 34*, 273–282.

Lee, S., & Trimi, S. (2016). Innovation for creating a smart future. *Journal of Innovation and Knowledge, 3*(1), 1–8. https://doi.org/10.1016/J.JIK.2016.11.001

Liker, J. K. (2004). *The Toyota way: 14 management principles from the world's greatest manufacturer*. New York, NY: McGraw-Hill.

Losonci, D., & Demeter, K. (2013). Lean production and business performance: International empirical results. *Competitiveness Review, 23*(3), 218–233.

Lucato, W. C., Vieira, M., Jr., Vanalle, R. M., & Salles, J. A. A. (2012). Model to measure the degree of competitiveness for auto parts manufacturing companies. *International Journal of Production Research, 50*(19), 5508–5522.

Macpherson, W., Lockhart, J., Kavan, H., & Iaquinto, A. (2015). Kaizen: A Japanese philosophy and system for business excellence. *Journal of Business Strategy, 36*(5), 3–9.

Meade, D. J., Kumar, S., & Houshyar, A. (2006). Financial analysis of a theoretical lean manufacturing implementation using hybrid simulation modeling. *Journal of Manufacturing Systems, 25*, 137–152.

Melton, T. (2005). The benefits of lean manufacturing: What lean thinking has to offer the process industries. *Chemical Engineering Research and Design, 83*(6), 662–673.

Moori, R. G., Pescarmona, A., & Kimura, H. (2013). Lean manufacturing and business performance in Brazilian firms. *Journal of Operations and Supply Chain Management, 6*(1), 91–105.

Moyano-Fuentes, J., & Sacristán-Díaz, M. (2012). Learning on lean: A review of thinking and research. *International Journal of Operations & Production Management, 32*, 551–582.

Myerson, P. (2012). *Lean supply chain and logistics management*. New York, NY: McGraw-Hill.

Nahm, A., Vonderembse, M., & Koufteros, X. (2003). The impact of organizational structure on time-based manufacturing and plant performance. *Journal of Operations Management, 21*(3), 281–306.

Ohno, T. (1988). *Toyota production system: Beyond large-scale production*. Cambridge, MA: Productivity Press.

Productivity Press Development Team. (1992). *Kaizen teian 2: Guiding continuous improvement through employee suggestions*. In Japan Human Relations Association (Ed.)., New York: Productivity Press.

Rahman, M. N. A., Khamis, N. K., Zain, R. M., Deros, B. M., & Mahmood, W. H. W. (2010). Implementation of 5S practices in the manufacturing companies: A case study. *American Journal of Applied Sciences, 7*(8), 1182–1189.

Raj, D., Ma, Y. J., Gam, H. J., & Banning, J. (2017). Implementation of lean production and environmental sustainability in the Indian apparel manufacturing industry: A way to reach the triple bottom line. *International Journal of Fashion Design, Technology and Education, 10*(3), 1–11.

Ramesh, V., & Kodali, R. A. (2012). A decision framework for maximising lean manufacturing performance. *International Journal of Production Research, 50*(8), 2234–2251.

Rondeau, P., Vonderembse, M., & Ragu-Nathan, T. (2000). Exploring work system practices for time-based manufacturers: Their impact on competitive capabilities. *Journal of Operations Management, 18*(5), 509–529.

Rose, A. M. N., Deros, B. M., Ab Rahman, M. N., & Nordin, N. (2011). Lean manufacturing best practices in SMEs. *Proceedings of the 2011 International Conference on Industrial Engineering and Operations Management, 2*(5), 872–877.

Rother, M. (2010). *Toyota kata: Managing people for improvement, adaptiveness, and superior results*. New York: McGraw-Hill.

Russell, R. S., & Taylor, B. W. (2000). *Operations management*. Englewood Cliffs, NJ: Prentice–Hall.

Sahoo, A. K., Singh, N. K., Shankar, R., & Tiwari, M. K. (2008). Lean philosophy: Implementation in a forging industry. *The International Journal of Advanced Manufacturing Technology, 36*, 451–462.

Shah, R., & Ward, P. T. (2003). Lean manufacturing: Context, practice bundles, and performance. *Journal of Operations Management, 21*(2), 129–149.

Shah, R., & Ward, P. T. (2007). Defining and developing measures of lean production. *Journal of Operations Management, 25*(4), 785–805.

Shaw, R., & Ward, P. T. (2003). Lean manufacturing: context, practice bundles, and performance. *Journal of Operations Management, 21*(2), 129–149.
Shewhart, W. A. (1939). *Statistical method from the viewpoint of quality control*. New York: Dover.
Shingo, S. (1989). *A study of the Toyota production system from an industrial engineering viewpoint*. Portland, OR: Productivity Press.
Shook, J., & Rother, M. (2003). *Aprendendo a Enxergar*. São Paulo: Editora Lean Institute Brasil.
Silva, E. M., Santos, F. C. A., & Castro, M. (2012). Analysis of the relationship between manufacturing strategy, practices and operational performance. *Production, 22*(3), 502–516.
Singh, H., & Mahmood, R. (2014). Manufacturing strategy and export performance of small and medium enterprises in Malaysia: Moderating role of external environment. *International Journal of Business and Commerce, 3*(5), 37–52.
Sukwadi, R., Wee, H., & Yang, C. (2013). Supply chain performance based on the lean-agile operations and supplier-firm partnership: An empirical study on the garment industry in Indonesia. *Journal of Small Business Management, 51*(2), 297–311.
Summers, D. C. S. (2009). *Quality management: Creating and sustaining organizational effectiveness* (2nd ed.). Upper Saddle River, NJ: Prentice Hall.
Swink, M., Narasinhan, R., & Kim, S. W. (2005). Manufacturing practices and strategy integration: Effects on cost efficiency, flexibility, and market-based performance. *Decision Sciences, 36*(3), 427–457.
Teece, D. (2007). Explicating dynamic capabilities: The nature and microfoundations of (sustainable) enterprise performance. *Strategic Management Journal, 28*(13), 1319–1350.
Terziovskia, M., & Sohalb, A. (2000). The adoption of continuous improvement and innovation strategies in Australian manufacturing firms. *Technovation, 20*(10), 539–550.
Timans, W., & Antony, J. (2012). Implementation of lean six sigma in small- and medium-sized manufacturing enterprises in the Netherlands. *Journal of the Operational Research Society, 63* (2), 339–353.
Treville, S., Shapiro, R. D., & Hameri, A. (2004). From supply chain to demand chain: The role of lead time reduction in improving demand chain performance. *Journal of Operations Management, 21*(6), 613–627.
Uma, D. S., & Kannan, A. (2010). Implementation of 5S in HRM. *Journal of Contemporary Research in Management, 5*(4).
Utiyama, M. H. R., & Godinho Filho, M. (2013). A literatura a respeito da comparação entre a teoria das restrições e a manufatura enxuta: Revisão, classificação e análise. *Revista Gestão E Produção, 20*(3), 615–638.
Votto, R. G., & Fernandes, F. C. F. (2014). Produção enxuta e teoria das restrições: Proposta de um método para implantação conjunta na indústria de bens de capital sob encomenda. *Revista Gestão E Produção, 21*(1), 45–63.
Wang, B. (2008). Analysis of efficiency of lean production implemented in multi-national optic enterprises. *International Journal of Technology Management, 43*(4), 304–319.
Westcott, R. T. (2006). *The certified manager of quality/organizational excellence handbook*. Milwaukee, WI: ASQ Quality Press.
Womack, J. P., & Jones, D. T. (2006). *Soluções enxutas: Como empresas e clientes conseguem juntos criar valor e riqueza*. Rio de Janeiro: Editora Campus.
Womack, J. P., & Jones, D. T. (2003). *Lean thinking: Banish waste and create wealth in your corporation*. London: Simon & Schuster UK Ltd.
Womack, J. P., Jones, D. T., & Ross, D. (1990). *The machine that changed the world*. New York, NY: Rawson Associates.
Worley, J. M., & Doolan, T. L. (2006). The role of communication and management support in a lean manufacturing implementation. *Management Decision, 44*(2), 228–245.
Wu, Y. C. (2003). Lean manufacturing: A perspective of lean suppliers. *International Journal of Operations and Production Management, 23*(11), 1349–1376.
Wyrwicka, M., & Mrugalska, B. (2017). Mirages of lean manufacturing in practice. *Procedia Engineering, 182*, 780–785.

Zakaria, N. H., Mohamed, N. M. Z. N., Ab Rahid, M. F. F., & Mohd, R. A. N. (2017). Lean manufacturing implementation in reducing waste for electronic assembly line. Proceeds of the *MATEC Web of Conferences* (Vol. 90, p. 01048). EDP Sciences.

Zayco, M. J., Broughman, D. J., & Hancock, W. M. (1997). Lean manufacturing yields world class improvements for small manufacturer. *IEE Solutions, 29*(4), 36–40.

Just in Time

3

3.1 The Foundations of the JIT System

In the present economy and in the face of a strong competitive economy, enterprises are looking for cost reduction and at the same time the manufacture of products of higher quality (Deuse et al., 2018). The progress of such enterprises is slowed down by limited resources. Thus, employing the lean manufacturing set of tools results in lower effort, less space occupied, lower lead times, fewer defective products and lower waste (Gandhi, Thanki, & Thakkar, 2017).

The concepts underpinning lean production are centred on the client/customer and can be reduced to five simple principles:

1. Solve completely the client's problem, assuring all requirements and working with the client.
2. Not wasting the client's time.
3. Delivering exactly *what* the client requests.
4. Delivering exactly what is required and where and *when*.
5. Aggregate permanently solutions to reduce the client's time and trouble/inconvenience.

In this regard, in any production system the ultimate goal is to generate profit. JIT is no exception. It is a market-oriented production system that aims to serve the customer's requirements (Botti, Mora, & Regattieri, 2017). In Hirano (2009), JIT is defined as an operations concept, which aims to meet requirements while offering perfect quality and zero waste. Thus, in each more competitive environment and as a minimal requirement for survival, manufacturing companies (not only, but in this book) should increase their capability to produce maximum quality products at the lowest cost and within the required time (Thomopoulos, 2016). However, this is not easy as it requires the commitment of all company collaborators and their strong focus on their own particular tasks (García-Alcaraz & Maldonado-Macías, 2016).

J. L. Q. Pinto et al., *Just in Time Factory*, Management for Professionals,
https://doi.org/10.1007/978-3-319-77016-1_3

There are several definitions of JIT to date. However, most of them focus on a way of thinking, the success of which depends on human factors, despite the inherent importance of machines (García-Alcaraz & Maldonado-Macías, 2016). All of these are acceptable, but García-Alcaraz and Maldonado-Macías (2016) chose to follow the simple adapted definition given by Vrat (2014): "JIT or zero-inventory system is an idealized concept of inventory management wherein we are able to supply whatever material is required, wherever required, and whenever required JIT with 100% supply assurances without keeping any inventory". The authors wished to add "with the highest possible quality", even if redundant.

According to Hirano (2009), from a comprehensive standpoint, a distinction between "JIT production management" and "JIT improvements" can be made. The first is a programme for developing production management adjusted to market needs and the second is a programme for improving efforts centred on a factory's physical system. The management system and the physical system must operate at the same rate, neither one pushing ahead nor holding back compared to the other. Otherwise, this is not the route to the company's success.

JIT is a management thinking tool that has been applied in practice in many Japanese manufacturing organizations. According to some authors, Toyota introduced JIT in the 1950s and continued to improve it (Dennis, 2015). Others argue it began in the early 1970s (Das & Handfield, 1997; Kootanaee, Nagendra Babu, & Talari, 2013). It gained extended support during the 1973 oil embargo and the increasing scarcity of other raw resources. In the following years, it was embraced by many other organizations. Toyota was capable of addressing the increasing challenges to its survival through a different management approach (Kootanaee et al., 2013). Later it was introduced in the US and other countries. The benefits of JIT have stimulated many Western companies to introduce this thinking or philosophy. This does not signify direct adoption: it is necessary to make appropriate changes according to the country and company culture and the failure to do so is one of reasons for various unsuccessful applications in the West. Many authors have presented several benefits deriving from JIT implementation (Heck & Zaidman, 2016; Zhou & Peng, 2017). However, not all companies achieved the same results and benefits as they were not operating in the same context or culture (García-Alcaraz & Maldonado-Macías, 2016; Schwerdfeger, Boysen, & Briskorn, 2017).

When an organization implements JIT production, the concept is generally associated with kanban (Michlowicz, 2017). This is not the most appropriate way to analyse JIT. Indeed, the vision is based on eliminating waste (muda), i.e. based on every action that does not add value to a product or service (Heiko, 1989). There are concerns that the consumption of resources and time makes products more expensive (Murray, 2006). According to Womack and Jones (2010), there are three types of activity in an industrial environment: (1) activity that adds value to the final product, which clearly should be maintained; (2) non-value-adding but unavoidable activity (muda Type I), which should be analysed and reduced whenever possible;

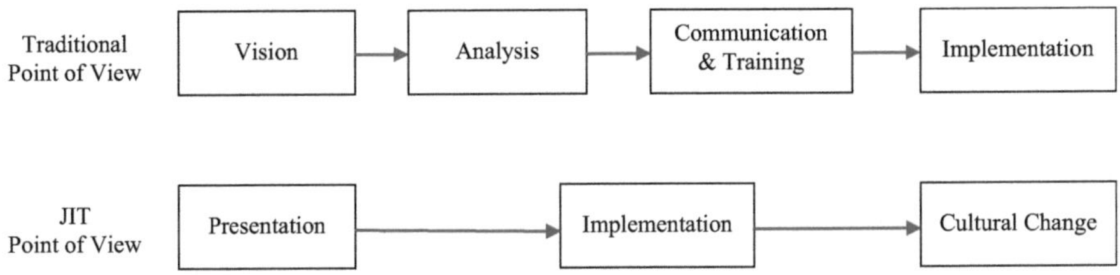

Fig. 3.1 Traditional versus JIT point of view

(3) non-value-adding and avoidable (muda Type II), which should/must be eliminated. Thus, the following measures must be adopted:

1. Execute only activities that create added value.
2. Ensure all activities executed create value from the client's viewpoint.
3. Ensure all steps in the implementation of a project have added value as an objective.
4. Produce only what is requested by the client.
5. Continuously and efficiently eliminate waste.

The JIT philosophy is based on the maximum reduction of the seven categories of waste in the production process, as described in 2.1.2. However, other types of waste should be considered, such as products and services that do not meet the client's requirements and take no advantage of employees' potential.

3.2 Production Management in JIT

3.2.1 JIT as a Tool

The JIT methodology includes a set of tools for the continuous analysis of the production process, such as kaizen, the pull system, 5S, takt-time and SMED, inter alia (Villa & Taurino, 2013). Through JIT application, the intention is to move from the idea to practice as quickly as possible, as shown in Fig. 3.1.

JIT is built on five basic assumptions, each of which is addressed in turn below.

1. Emphasis on creating added value in production and information flows

A complete analysis of the product manufacturing time should be carried out, from the client's request until delivery. This is defined as the lead time and the analysis aims to achieve the overall optimization of available resources. The intention is to do more with less and less, reducing waste and approaching always what clients want in such a way that it creates value for them, as shown in Fig. 3.2.

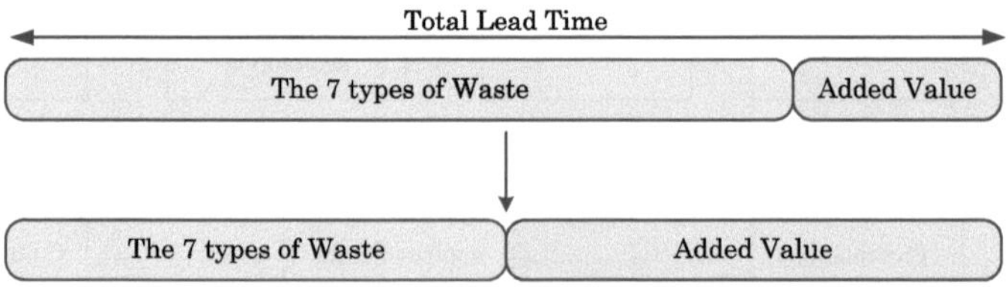

Fig. 3.2 Focus on added value creation

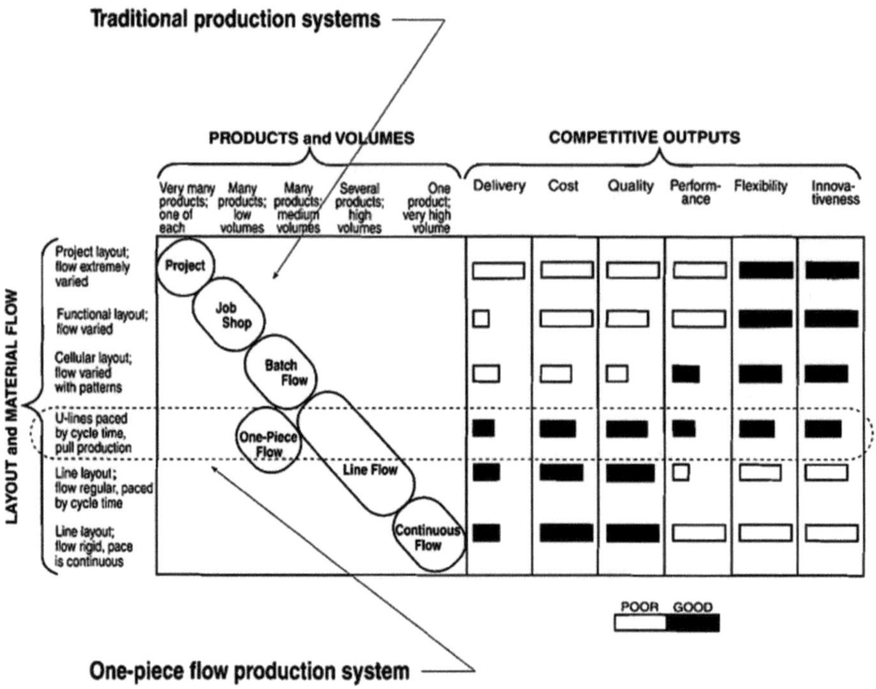

Fig. 3.3 Traditional batch flow. Source: Miltenburg (2001)

2. Systematic leadership with an entrepreneurial vision

There should be leadership in place enabling in-depth understanding of client requirements, thus generating satisfaction based on such requirements. This leadership should be experienced, with a global view of the full production chain. It should also generate profitability, be able to manage crisis situations and offer technical support whenever necessary. Leadership should moreover be versatile, enabling it to act appropriately to exigencies.

3. Continuous and cadenced pulled flow

During product and/or process development, the information flow and knowledge should be available to all employees (at the exact time, in the exact place and in the required amount). Figure 3.3 shows the differences between several production systems. For example, in traditional batch systems there are separate workstations, intermediate inventories and a lack of information sharing among operators. In one piece flow systems (lean production) there are operators who are linked and no intermediate inventory, only components and information throughout the production chain and work piece by piece. This information is very important and allows the sustained development of new products and/or processes. Information flow and knowledge should maintain a continuous rhythm. In this context, employees should be able to pull this information and acquire knowledge at an in-depth level.

4. Specialists' responsibility

During the development of a product and/or process, the information flow should involve all those involved in the project. Specialists are responsible for overcoming technological barriers and finding a profitable means of production and information flow with maximum value added. When placing a new product and/or process in the continuous manufacturing phase, all information should be provided to the next players.

5. Concurrent or simultaneous engineering

The TPS proposes that the various possible departments be involved whenever a new project and/or process is initiated. The main goal is to provide an adequate analysis of all alternatives in order to prevent a product/process project failing to be compliant.

In this regard, all those involved in the process know their functions in the future, thus enabling them to address bottlenecks and priority actions, among others. Thus, they can initiate contacts with suppliers to inform them what requirements should be respected. Also, they can work with the clients to inform them of the project's development state.

3.2.2 JIT and Production Planning

Production planning aims to create an efficient production system in the following four vectors:

- Quality—It is necessary to determine the real production process capacities, verify if the specifications are adequate, work always on process development, equipment and people and ensure the product quality required.

- Cost—As noted previously, the essence of JIT methodology is the focus on eliminating any type of waste, known in Toyota as a "non-cost principle". According to this perspective, the price should not be *Cost* + *Profit* = *Price*, but rather *Profit* = *Price* − *Cost*.
- Employees—The JIT philosophy aims for employees to be as flexible as possible, facilitating their adaptation to any changes that may arise in the product manufacturing and/or in technology.
- Delivery—The products must be produced according to the customer's requirements and should be delivered at the required time.

3.2.3 JIT and the Human Dimension

The traditional approach to employees' work changes in the JIT environment. Employees execute more challenging tasks as more responsibility is imbued on an organizational basis (Uhrin, Bruque-Cámara, & Moyano-Fuentes, 2017). On the other hand, employees may incur more stress in their work (Conti, Angelis, Cooper, Faragher, & Gill, 2006).

In JIT, human work is characterized by the following aspects:

- Work is understood and executed at the workplace.
- There is decentralization in decision making.
- There is a less vertical hierarchy.
- There is direct collaboration between engineers and workers.
- There is collaboration between maintenance and production (transfer of responsibilities).
- There is interaction between the design and development department and the industrial units.

Boyer (1996) considered that the existence of the four indicators measure the degree of organization and employees' commitment to this thinking. These include: leadership with regard to quality, the establishment of improvement groups, continuous training and the delegation of responsibility to employees. However, according to Kochan, Lansbury, and MacDuffie (1997), in analysing the application of the JIT model in the automobile industry, there has been a mismatch between its use in countries such as the US and Canada and in Japan and Europe. In the US and Canada only 23% and 4% of the employees respectively were found to be integrated in the work teams, whereas in Japan and Europe the levels of involvement reached 70–80%. These data reveal the attitudes of the workers' syndicates in North American countries, which do not allow the diffusion of this model in the industry concerned. In this regard, it is possible to conclude that this model can be applied when companies, syndicates and governmental strategies are aligned. Thus, there is a substantial relationship between the successful implementation of the JIT model and the employees' identification with the strategy followed. Figures 3.4 and 3.5 show the importance of the employees in the TPS structure.

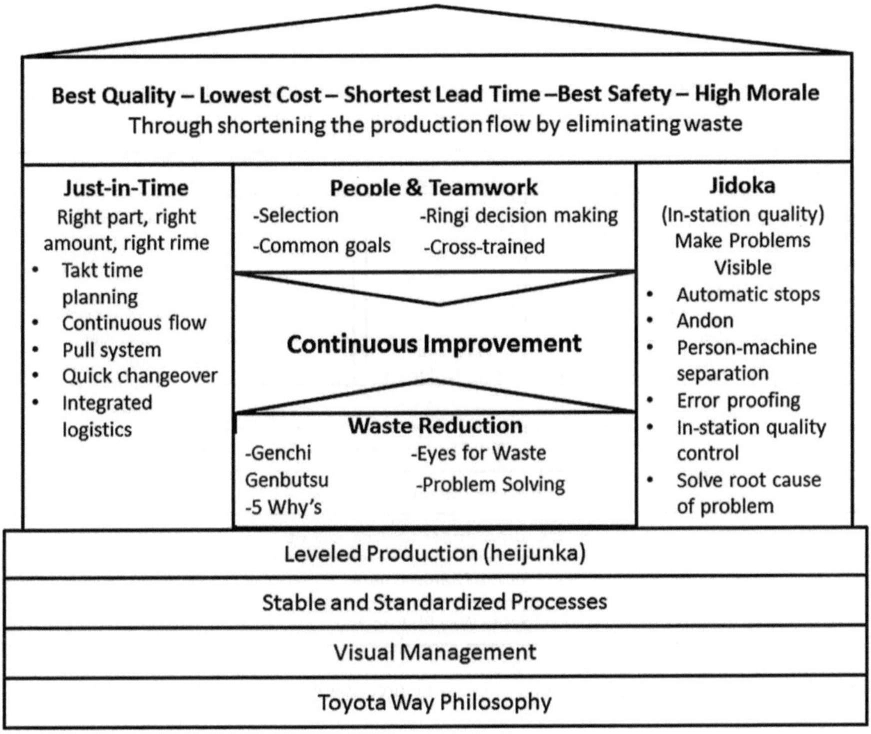

Fig. 3.4 Classical Toyota production system structure. Source: Kehr and Proctor (2017)

The full utilization of workers' capabilities is the second basic concept of Toyota (Towill, 2010). Toyota built a system ensuring respect for humans. This system is based on the following points: (1) elimination of workers' wasted movements; (2) consideration for workers' occupational health and safety (OHS); (3) self-display of workers' capabilities by delegating them greater responsibility and authority. Workers take part in making improvements. Each of Toyota's employees has a right to make an improvement in terms of waste found. Thus, in JIT production, all processes and all plants are oriented to no surplus. If a problem remains, the line will immediately stop running and this will affect the entire plant. The need for improvement is easily understood by all. The authority and responsibility for improving workstations is delegated to the workers themselves, which is the most distinctive feature of the JIT mode of working and an example of Toyota's respect for the human dimension of production (Vaszkun & Tsutsui, 2012).

García-Alcaraz and Maldonado-Macías (2016) consider that if JIT is considered a production philosophy, some of its elements must be associated with human resources or factors, as any philosophy is inherently related to humans. The authors present a list

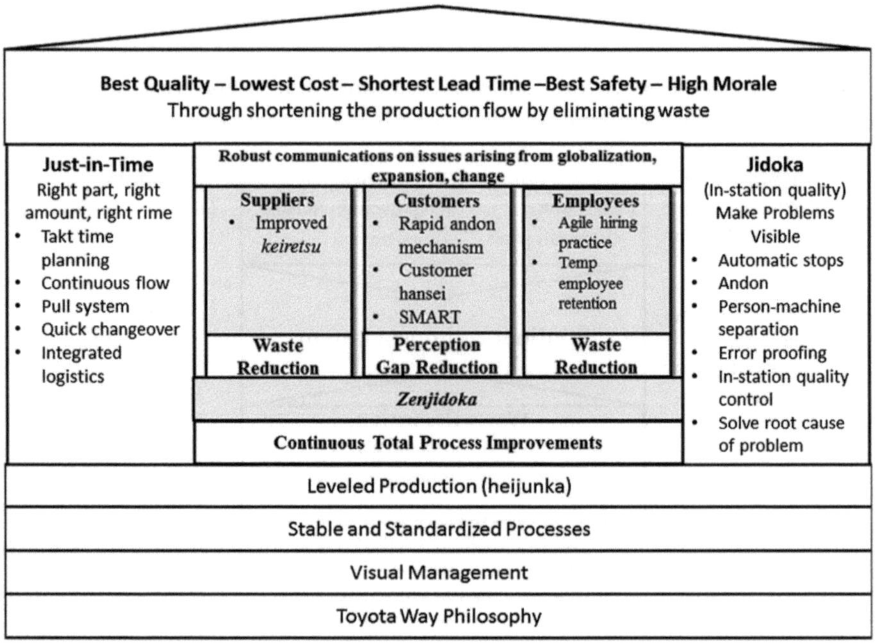

Fig. 3.5 Expanded Toyota production system structure. Source: Kehr and Proctor (2017)

from a literature review that shows JIT elements associated with human resources (García-Alcaraz & Maldonado-Macías, 2016; Uhrin et al., 2017; Wickramasinghe & Wickramasinghe, 2016):

- Flexible workforce
- Zero deviation in production programmes
- Multifunctional workers
- Worker motivation
- Short delivery times
- Prevention of errors
- Long-term contracts
- Self-correction of errors
- Employee empowerment in quality control
- Quality certification of suppliers
- Suppliers' evaluation and selection
- Effective communication

This is the full demonstration of employees' importance in the process of JIT implementation. Moreover, Kehr and Proctor (2017) identify potential structural shortcomings and possible deficiencies of the TPS and suggest potential re-structuring to achieve more optimal continuous improvement based on the root causes of recall crises (see Figs. 3.4 and 3.5. In their new expanded TPS structure (Fig. 3.5), compared to the traditional TPS structure (Fig. 3.4), an "employee" pillar is proposed to capture states, transitions and turbulence among employees that result from increased globalization, expansion and change. With this in mind, organizations should implement a company-wide skill-mapping system. The aim is to derive quality assurance-related skills and adopt effective metrics for the evaluation of these skills and to motivate employees to hone and apply them (Kehr & Proctor, 2017).

3.2.4 The Role of JIT in the TPS Structure

It is well known that JIT has the ability to reduce costs by reducing and eliminating waste. However, the TPS should not be interpreted as essentially JIT methodology. Rather, JIT is only one of the pillars. For this to be considered the only pillar would be to limit its scope and potentiality. Instead, JIT and jidoka are two methodologies that together create the stability of the TPS (Jadhav, Mantha, & Rane, 2015; Vörös & Rappai, 2016).

Initially, researchers (Monden, 2011; Pettersen, 2009) focused only on the JIT concept due to its capability to reduce inventory and because of its tangible benefits. The best characteristic of JIT is its aim to ensure the fulfilment of product requirements at the right time, in the right amount and eliminating unnecessary stocks (Chen & Tan, 2013). TPS concerns not only the elimination of non-added-value activities from the process. It is also related to improving product quality, aided by jidoka. The success of the TPS has resulted in its wide acceptance by manufacturing industries globally. It has since been disseminated into other non-conventional industries. The TPS only preceded the foundation of lean production (Jasti & Kodali, 2015).

To support JIT, Ohno and Bodek (1988) developed the Kanban concept. Jidoka (also called autonomation) encompasses the notion of automation including an operator's role (Nadal, 2017). The goal is not to eliminate workers, but rather to concentrate their work on aspects of higher valued practice (Chen, 2015). As used in Toyota, jidoka means ensuring the equipment or operations stop whenever a non-standard or defective condition arises (Rüttimann, 2018). If trouble with equipment or a machine defect occurs, the equipment or entire line stops and any line with workers can be stopped by them. The importance of jidoka is that it motivates workers by preventing them from making too much and easing deviation control. To make improvements, it is only necessary to direct attention to the stopped equipment and the worker who activated the stoppage. This is a central requirement when addressing the full utilization of workers' capabilities in constructing the system (Arredondo-Soto, Carrillo-Gutiérrez, Solís-Quinteros, & Hernández-Escobedo, 2018).

There are several ways of representing the TPS structure. Figures 3.4 and 3.5 provide schematic presentations of the essentials of the production system and the pillars on which this system is based. JIT and jidoka. According to this model,

Toyota's purpose is to meet customer requirements, providing products and services to meet their expectations, at the lowest cost and in the shortest possible time (lead time). However, it must not be forgotten that it is necessary to provide good conditions for employees in the form of a good working environment and good OHS conditions, whereby ethical standards concerning workers are guaranteed. Kehr and Proctor (2017) identified potential structural inadequacies and possible TPS deficiencies, bearing in mind the root causes of recall crises. They suggested potential restructuring to achieve better continuous improvement (Figs. 3.4 and 3.5). In this new expanded TPS structure (Fig. 3.5) and compared to the traditional TPS structure (Fig. 3.4), an "employee" pillar is proposed for apprehending states, transitions and instability among employees as a consequence of wider globalization, development and change. To address this, organizations should develop and implement company-wide skill mapping for quality assurance related to skills, after adopting useful metrics for the evaluation of these skills. The final aim is to motivate employees to refine and apply them (Kehr & Proctor, 2017). JIT and jidoka are methodologies that together create the stability of TPS. TPS is mostly governed by "pull" production, based on JIT process control, jidoka and flexible workforce utilization.

3.2.5 Impacts of JIT

Pull production was mainly popularized by the adoption of JIT principles. Recent works have shown that JIT is nowadays able to promote the proficient, effective and flexible use of productive resources, reducing lead times, decreasing throughput times (Chen & Sarker, 2017) and reducing inventories (Zhang, 2014), despite some doubts concerning the benefits of inventory reduction (Folinas, Fotiadis, & Coudounaris, 2017).

Mackelprang and Nair (2010) conducted an significant empirical study regarding the impact of JIT practices on performance. Bortolotti, Danese, and Romano (2013) studied the relationship between JIT manufacturing practices and performance using a sample of 25 articles and determined that the connexion between JIT and performance is influenced majority by "moderating" plant-specific factors such as plant age, production characteristics, plant size, geographic location, organizational structure and culture, among others. But they suggest that a gap in the literature exists in studying that relationship based on a list of contextual variables (Godinho Filho, Ganga, & Gunasekaran, 2016).

By means of a survey of US plants, Mehra and Inman (1992) identified the elements of JIT implementation to ensure fruitful implementation. They conclude that the success of JIT implementation is associated with high levels of a JIT production strategy and JIT vendor strategy. In turn, Sakakibara, Flynn, and Schroeder (1993) identified equipment layout, pull system support, supplier quality level and a Kanban system as the major contributing elements of JIT influencing performance in the companies studied (41 plants in Japan and the US). Lawrence and Hottenstein (1995) found a clear positive relationship between JIT and performance using 124 plants surveyed in Mexico. They also identified plant size, industry and type of production process as variables that influence the relationship between JIT practices and performance. Callen, Fader, and Krinsky (2000) concluded that JIT practices are associated with improved inventory performance and higher profitability based on a survey of 100 Canadian plants (Godinho Filho et al., 2016).

Fullerton and McWatters (2001) found a considerable relationship between JIT practices and lower inventory levels, improved quality performance and responsiveness and financial returns. They also exhibited that larger firms are more probable to exhibit comprehensive JIT implementation and this leads to better overall performance. Shah and Ward (2003) provided support for the influence of plant size on lean implementation and performance. They found a significant and positive relationship between lean implementation and operational performance, investigating the implementation of 22 lean manufacturing practices in the US (Godinho Filho et al., 2016).

McKone, Schroeder, and Cua (2001) provided evidence that JIT practices should be supported by productive maintenance efforts. These together improve manufacturing performance. Other studies have shown that JIT/lean can be associated with other manufacturing technologies/paradigms, such as advanced manufacturing technologies (Das & Jayaram, 2003), agile manufacturing (Inman, Sale, Green, & Whitten, 2011) and environmental management (Yang, Hong, & Modi, 2011) (Godinho Filho et al., 2016).

Hou, Chan, and Wang (2011) presented key findings concerning the implementation of JIT systems according to five themes (Information Systems, Production Planning, Inventory Management, Quality Management, Supplier Management) in a case study of the Chinese automotive industry. This research corroborates the benefits of applying JIT systems as advocated in many studies, in particular in terms of logistics and relationships with suppliers, which are of vital importance, despite the single case study employed. Qureshi, Iftikhar, Bhatti, Shams, and Zaman (2013) studied factors which have a positive relationship with JIT implementation in the cement industry of Pakistan. This study focused on product design, total quality control, inventory management, supply chain integration and production planning and the link to JIT implementation. JIT implementation was measured through six indicators: overproduction, waiting time, transportation, inappropriate processing, unnecessary motion and defects. This study found evidence of a positive relationship between JIT implementation and product design, a direct relationship between TQC and JIT management implementation, a direct relationship between JIT and inventory management and planning (implementing JIT reduced unnecessary inventory), a positive relationship between supply chain integration and JIT implementation and a positive relationship between product planning and JIT.

References

Arredondo-Soto, K. C., Carrillo-Gutiérrez, T., Solís-Quinteros, M., & Hernández-Escobedo, G. (2018). A theoretical framework about the impact of human factors on manufacturing process performance. In J. L. García-Alcaraz, G. Alor-Hernández, A. A. Maldonado-Macías, & C. Sánchez-Ramírez (Eds.), *New perspectives on applied industrial tools and techniques, management and industrial engineering* (pp. 327–352). Cham: Springer.

Bortolotti, T., Danese, P., & Romano, P. (2013). Assessing the impact of just-in-time on operational performance at varying degrees of repetitiveness. *International Journal of Production Research, 51*, 1117–1130. https://doi.org/10.1080/00207543.2012.678403

Botti, L., Mora, C., & Regattieri, A. (2017). Integrating ergonomics and lean manufacturing principles in a hybrid assembly line. *Computers & Industrial Engineering, 111*, 481–491. https://doi.org/10.1016/j.cie.2017.05.011

Boyer, K. K. (1996). An assessment of managerial commitment to lean production. *International Journal of Operations & Production Management, 16*, 48–59. https://doi.org/10.1108/01443579610125589

Callen, J. L., Fader, C., & Krinsky, I. (2000). Just-in-time: A cross-sectional plant analysis. *International Journal of Production Economics, 63*, 277–301. https://doi.org/10.1016/S0925-5273(99)00025-0

Chen, Z. (2015). The relationships among JIT, TQM and production operations performance: An empirical study from Chinese manufacturing firms. *Business Process Management Journal, 21*, 1015–1039. https://doi.org/10.1108/BPMJ-09-2014-0084

Chen, Z., & Sarker, B. R. (2017). Integrated production-inventory and pricing decisions for a single-manufacturer multi-retailer system of deteriorating items under JIT delivery policy. *International Journal of Advanced Manufacturing Technology, 89*, 2099–2117. https://doi.org/10.1007/s00170-016-9169-0

Chen, Z., & Tan, K. H. (2013). The impact of organization ownership structure on JIT implementation and production operations performance. *International Journal of Operations & Production Management, 33*, 1202–1229. https://doi.org/10.1108/IJOPM-06-2011-0207

Conti, R., Angelis, J., Cooper, C., Faragher, B., & Gill, C. (2006). The effects of lean production on worker job stress. *International Journal of Operations & Production Management, 26*, 1013–1038. https://doi.org/10.1108/01443570610682616

Das, A., & Handfield, R. B. (1997). Just-in-time and logistics in global sourcing: An empirical study. *International Journal of Physical Distribution & Logistics Management, 27*, 244–259. https://doi.org/10.1108/09600039710170601

Das, A., & Jayaram, J. (2003). Relative importance of contingency variables for advanced manufacturing technology. *International Journal of Production Research, 41*, 4429–4452. https://doi.org/10.1080/00207540310001595819

Dennis, P. (2015). *Lean production simplified: A plain-language guide to the world's most powerful production system* (3rd ed.). Boca Raton, FL: Productivity Press.

Deuse, J., Heuser, C., Konrad, B., Lenze, D., Maschek, T., Wiegand, M., & Willats, P. (2018). Pushing the limits of lean thinking – Design and management of complex production systems. In E. Viles, M. Ormazábal, & A. Lleó (Eds.), *Closing the gap between practice and research in industrial engineering* (pp. 335–342). Cham: Springer.

Folinas, D. K., Fotiadis, T. A., & Coudounaris, D. N. (2017). Just-in-time theory: The panacea to the business success? *International Journal of Value Chain Management, 8*, 171–190. https://doi.org/10.1504/IJVCM.2017.085485

Fullerton, R. R., & McWatters, C. S. (2001). The production performance benefits from JIT implementation. *Journal of Operations Management, 19*, 81–96. https://doi.org/10.1016/S0272-6963(00)00051-6

Gandhi, N. S., Thanki, S. J., & Thakkar, J. J. (2017). Ranking of drivers for integrated lean-green manufacturing for Indian manufacturing SMEs. *Journal of Cleaner Production, 171*, 675–689. https://doi.org/10.1016/j.jclepro.2017.10.041

García-Alcaraz, J. L., & Maldonado-Macías, A. A. (2016). *Just-in-time elements and benefits, management and industrial engineering*. Cham: Springer.
Godinho Filho, M., Ganga, G. M. D., & Gunasekaran, A. (2016). Lean manufacturing in Brazilian small and medium enterprises: implementation and effect on performance. *International Journal of Production Research, 54*(24), 7523–7545. https://doi.org/10.1080/00207543.2016.1201606
Heck, P., & Zaidman, A. (2016). A framework for quality assessment of just-in-time requirements: The case of open source feature requests. *Requirements Engineering, 22*, 1–21. https://doi.org/10.1007/s00766-016-0247-5
Heiko, L. (1989). A simple framework for understanding JIT. *Product Inventory Management Journal, Alexandria, 30*, 61.
Hirano, H. (2009). *JIT implementation manual: The complete guide to just-in-time manufacturing* (2nd ed.). New York: Taylor and Francis.
Hou, B., Chan, H. K., & Wang, X. (2011). A case study of just-in-time system in the Chinese automotive industry. In the *Proceedings of the World Congress of Engineers 2011 (The 2011 International Conference of Manufacturing Engineering and Engineering Management)* (pp. 904–908).
Inman, R. A., Sale, R. S., Green, K. W., & Whitten, D. (2011). Agile manufacturing: Relation to JIT, operational performance and firm performance. *Journal of Operations Management, 29*, 343–355. https://doi.org/10.1016/j.jom.2010.06.001
Jadhav, J. R., Mantha, S. S., & Rane, S. B. (2015). Analysis of interactions among the barriers to JIT production: Interpretive structural modelling approach. *Journal of Industrial Engineering International, 11*, 331–352. https://doi.org/10.1007/s40092-014-0092-4
Jasti, N. V. K., & Kodali, R. (2015). A critical review of lean supply chain management frameworks: Proposed framework. *Production Planning & Control, 26*, 1051–1068. https://doi.org/10.1080/09537287.2015.1004563
Kehr, T. W., & Proctor, M. D. (2017). People pillars: Re-structuring the Toyota production system (TPS) house based on inadequacies revealed during the automotive recall crisis. *Quality and Reliability Engineering International, 33*, 921–930. https://doi.org/10.1002/qre.2059
Kochan, T. A., Lansbury, R. D., & MacDuffie, J. P. (Eds.). (1997). *After lean production: Evolving employment practices in the world auto industry. Cornell International Industrial and Labor Relations Reports*. Ithaca, NY: Cornell University Press.
Kootanaee, A. J., Nagendra Babu, K., & Talari, H. F. (2013). Just-in-time manufacturing system: From introduction to implement. *International Journal of Economics, Business and Finance, 1*(2), 07–25.
Lawrence, J. J., & Hottenstein, M. P. (1995). The relationship between JIT manufacturing and performance in Mexican plants affiliated with U.S. companies. *Journal of Operations Management, 13*, 3–18. https://doi.org/10.1016/0272-6963(95)00018-N
Mackelprang, A. W., & Nair, A. (2010). Relationship between just-in-time manufacturing practices and performance: A meta-analytic investigation. *Journal of Operations Management, 28*, 283–302. https://doi.org/10.1016/j.jom.2009.10.002
McKone, K. E., Schroeder, R. G., & Cua, K. O. (2001). The impact of total productive maintenance practices on manufacturing performance. *Journal of Operations Management, 19*, 39–58. https://doi.org/10.1016/S0272-6963(00)00030-9
Mehra, S., & Inman, R. A. (1992). Determining the critical elements of just-in-time implementation. *Decision Science, 23*, 160–174. https://doi.org/10.1111/j.1540-5915.1992.tb00382.x
Michlowicz, E. (2017). Kanban system in the flow control subassemblies as a component of lean manufacturing. In proceeding of *Intelligent Systems in Production Engineering and Maintenance – ISPEM 2017, Advances in intelligent systems and computing* (pp. 271–283). Cham: Springer. https://doi.org/10.1007/978-3-319-64465-3_27
Miltenburg, J. (2001). U-shaped production lines: A review of theory and practice. *International Journal of Production Economics, 70*(3), 201–214.
Monden, Y. (2011). *Toyota production system: An integrated approach to just-in-time* (4th ed.). Boca Raton, FL: Productivity Press.

Murray, C. (Ed.). (2006). *The marketing gurus: Lessons from the best marketing books of all time*. New Yok: Penguin.

Nadal, J. O. (2017). Lean management and supply chain management: Common practices. In A. P. Barbosa Póvoa, A. Corominas, J. L. de Miranda (Eds.), Optimization and decision support systems for supply chains (pp. 117–129). Cham: Springer.

Ohno, T., & Bodek, N. (1988). *Toyota production system: Beyond large-scale production* (1st ed.). Cambridge, MA: Productivity Press.

Pettersen, J. (2009). Defining lean production: Some conceptual and practical issues. *Total Quality Management Journal, 21*, 127–142. https://doi.org/10.1108/17542730910938137

Qureshi, M. I., Iftikhar, M., Bhatti, M. N., Shams, T., & Zaman, K. (2013). Critical elements in implementations of just-in-time management: Empirical study of cement industry in Pakistan. *SpringerPlus, 2*, 645. https://doi.org/10.1186/2193-1801-2-645

Rüttimann, B. G. (2018). *Lean compendium*. Cham: Springer.

Sakakibara, S., Flynn, B. B., & Schroeder, R. G. (1993). A framework and measurement instrument for just-in-time manufacturing. *Production Operations Management, 2*, 177–194. https://doi.org/10.1111/j.1937-5956.1993.tb00097.x

Schwerdfeger, S., Boysen, N., & Briskorn, D. (2017). Just-in-time logistics for far-distant suppliers: Scheduling truck departures from an intermediate cross-docking terminal. *OR Spectrum, 40*, 1–21. https://doi.org/10.1007/s00291-017-0486-y

Shah, R., & Ward, P. T. (2003). Lean manufacturing: Context, practice bundles, and performance. *Journal of Operations Management, 21*, 129–149. https://doi.org/10.1016/S0272-6963(02)00108-0

Thomopoulos, N. T. (2016). *Elements of manufacturing, distribution and logistics*. Cham: Springer.

Towill, D. R. (2010). Industrial engineering the Toyota production system. *Journal of Management History, 16*, 327–345. https://doi.org/10.1108/17511341011051234

Uhrin, Á., Bruque-Cámara, S., & Moyano-Fuentes, J. (2017). Lean production, workforce development and operational performance. *Management Decision, 55*, 103–118. https://doi.org/10.1108/MD-05-2016-0281

Vaszkun, B., & Tsutsui, W. M. (2012). A modern history of Japanese management thought. *Journal of Management History, 18*, 368–385. https://doi.org/10.1108/17511341211258729

Villa, A., & Taurino, T. (2013). From JIT to Seru, for a production as lean as possible. *Procedia Engineering, The Manufacturing Engineering Society International Conference, MESIC 2013, 63*, 956–965. https://doi.org/10.1016/j.proeng.2013.08.172

Vörös, J., & Rappai, G. (2016). Process quality adjusted lot sizing and marketing interface in JIT environment. *Applied Mathematical Modelling, 40*, 6708–6724. https://doi.org/10.1016/j.apm.2016.02.011

Vrat, P. (2014). *Materials management*. New Delhi: Springer.

Wickramasinghe, V., & Wickramasinghe, G. L. D. (2016). Variable pay and job performance of shop-floor workers in lean production. *Journal of Manufacturing Technology Management, 27*, 287–311. https://doi.org/10.1108/JMTM-12-2014-0130

Womack, J. P., & Jones, D. T. (2010). *Lean thinking: Banish waste and create wealth in your corporation*. New York: Simon and Schuster.

Yang, M. G., Hong, P., & Modi, S. B. (2011). Impact of lean manufacturing and environmental management on business performance: An empirical study of manufacturing firms. *International Journal of Production Economics, 129*, 251–261. https://doi.org/10.1016/j.ijpe.2010.10.017

Zhang, X. J. (2014). Inventory control of supply chain environment. *Advanced Material Research, 971–973*, 2346–2349. https://doi.org/10.4028/www.scientific.net/AMR.971-973.2346

Zhou, B., & Peng, T. (2017). Scheduling the in-house logistics distribution for automotive assembly lines with just-in-time principles. *Assembly Automation, 37*, 51–63. https://doi.org/10.1108/AA-04-2016-028

Lean Manufacturing Tools

4

Successful implementation of lean manufacturing depends on the maximum involvement of personnel and the application of a set of tools or frameworks such as 5S, SMED, total productive maintenance (TPM) and Kanban (Chiarini & Vagnoni, 2015).

Lean manufacturing is a methodology that aims to minimize resource consumption and waste in enterprise activities (Luo, Mo, & Ruo, 2007; Phogat, 2013). The last two decades have seen the rise of several lean production theories and methodologies, such as the combination of mass customization and lean production, cell production, JIT, 5S and TPM. Many companies around the world have applied lean principles and methods to production development and have created management systems attending to their needs and customers' requirements (Liker & Morgan, 2011; Nepal, Yadav, & Solanki, 2011).

Lean manufacturing suggests several tools to help companies reduce waste. The lean manufacturing' tools with higher levels of implementation among manufacturing companies are: value stream mapping (VSM) (Rother & Shook, 2003a, 2003b), 5S (Brunet, 2003), cellular manufacturing (Ohno, 1988a, 1988b, 1988c), SMED (Shingo, 1989) and TPM (Monden, 1998a, 1998b, 1998c). These and other lean manufacturing tools are identified and clustered according to their application context in Table 4.1.

4.1 The 5S: Seiri, Seiton, Seisou, Seiketsu, Shitsuke

5S stands for *seiri*, *seiton*, *seisou*, *seiketsu* and *shitsuke* in Japanese, which are often translated into English as sort, set in order, shine, standardize and sustain. Sometimes other terms are used, but they mean the same thing. Traditionally, 5S is used to organize physical space and items (workspace, desk, charts, department and so on) (Tapping, Kozlowski, Archbold, & Sperl, 2009) through the elimination of waste, providing visual control and preparing the workplace to enable improvements to be effective and implemented efficiently. The concept of 5S has

J. L. Q. Pinto et al., *Just in Time Factory*, Management for Professionals,
https://doi.org/10.1007/978-3-319-77016-1_4

Table 4.1 Types of lean tools

Type	Lean tools
Machinery and equipment	Low-cost automation
	Overall equipment effectiveness (OEE)
	Preventive maintenance
	Single-minute exchange of dies (SMED)
	Total productive maintenance (TPM)
Material flow and layout	Cellular manufacturing
	First in first out (FIFO)
	One-piece-flow
	Simulation software
	Optimization of the supply chain
	Value stream mapping (VSM)
	Workstation design
Organization and staff	5S
	Autonomous work groups
	Benchmarking
	Ideas management
	Job rotation
	Lean office (Administration)
	Kaizen
	Standardization
Quality	FMEA
	Poka Yoke
	Quality circles

Source: Zakaria, Mohamed, Ab Rahid, and Rose (2017)

its origins in socio-historical and philosophical fields (Kobayashi, 2005) in Japan, consistent with "Japanese thinking", which is a balance between philosophy and technique and is used to coach the body and mind through discipline (Sugiura & Gillespiere, 2002).

The lean tool 5S was first introduced together with TPM in the late 1980s and early 1990s (Nakajima, 1988; Takahashi & Osada, 1990). 5S represents a set of practices for improving workplace organization and productivity (Gapp, Fisher, & Kobayashi, 2008) contributing to maximizing value added by removing all non value-added factors (Womak, Jones, & Roos, 1990). Initially, it was mainly implemented in the industrial sector, but it has recently been applied to the services sector as a systematic method of organizing and standardizing the workplace (Jackson, 2009). It is widely implemented around the world based on its low-cost implementation and technologically undemanding approach and the fact that it is considered a starting point for the improvement of organizations (Imai, 2012).

The 5S is viewed differently in different geographical regions. This was observed by Kobayashi, Fisher, and Gapp (2008). In Japan the implementation of 5S is viewed as a strategy for organizational excellence, but in the United Kingdom (UK) and US it is viewed as a tool solely for workplace organization. In other countries, the implementation of 5S is viewed as a minimum requirement for health and safety in the workplace which has led to extension of the scope of 5S to incorporate a new S, namely "safety and health" (Zelinski, 2005). Incorrect interpretation and implementation of this lean tool has also been focused on in some studies. Lixia and Bo (2011) highlighted the main problems faced by Chinese companies in implementing 5S, proposing a set of steps to a successful implementation of a 5S programme.

Beyond the geographical context, there are other factors influencing 5S implementation and system performance after implementing it, such as firm size, product type, organizational attitude towards quality and continuous improvement (Bayo-Moriones, Bello-Pintado, & Merino-Díaz de Cerio, 2010). Also, Brown and O'Rourke (2007) highlight employee engagement through training and involvement and Ho (1999) note the need to involve the whole organization to ensure complete and systematic implementation of the 5S tool.

The 5S lean tool can be summarized as follows (Ramdass, 2015):

Seiri—Sort. This refers to the elimination of unnecessary materials, tools, equipment and furniture from the workplace. A criterion that could be used to identify these is frequency of use. This step results in free floor space, better product flow and improved communication.

Seiton—Set in order (straighten). In this step, goals are set to ensure that everything is kept in its place. It concerns the arrangement, placement and sequence of items left after the "sort" step. This step contributes to the elimination of numerous types of waste, improved quality and reduction of inventory levels.

Seiso—Shine. This step is associated to the cleanness of the workplace. It requires the construction of a 5S job cycle chart listing the 5S tasks, the pillar each represents and the frequency with which the task is to be performed. This step is quite important contributing to increasing individual pride in the work area, creating a safer workplace and reducing defects and equipment downtime.

Seiketsu—Standardize. This consists of the standardization of the process of maintaining the previous steps by dedicating a short period of time in every day to cleaning up in order to prevent excess inventory and unwanted items in the workplace.

Shitsuke—Sustain. To form a habit of continuous improvement, this step implements frequent auditing procedures in each area in which 5S is implemented, or entails adopting a programme in which a group of individuals in conjunction with the team leader visit other areas on the shop floor.

5S represents a simple and non-costly tool to achieve the operational benefits of lean manufacturing. Considerable effort has been dedicated to listing ways of how to implement it, even to the extent of detailing day-to-day activities on the shop floor.

4.1.1 Requirements for 5S Implementation

The review of the literature has revealed a wide range of factors that are strategic for the successful implementation of lean and associated tools, summarized by Salonitis and Tsinopoulos (2016) as follows:

- Organizational culture and ownership
- Developing organizational readiness
- Management commitment and capabilities
- Providing adequate resources to support change
- External support from consultants in the first instance
- Effective communication and engagement
- Strategic approach to improvements
- Teamwork and joined-up whole systems thinking
- Timing to set realistic timescales for change and to make effective use of commitment to and enthusiasm for change

According to Lockman (2015), there is a set of factors that determine the success of 5S implementation:

1. The right timing—The first three Ss should be implemented in a three-month window. It is essential to provide a workable implementation deadline.
2. No-nonsense practicality—5S is not only very rational, which makes it critical to provide training courses or manuals for employees but also their tasks are very hands-on.
3. A healthy serving of spaghetti (charts)—When implementing a 5S approach, it is crucial always to examine different configurations of operations first. To perform this examination, a spaghetti chart should be created to capture motion.
4. Using all 5Ss—No matter how well-cleaned and organized the company's 5S operations are, they will quickly revert to a state of disarray if the final S is neglected.

A suggested approach to implementing 5S in an effective way is to use the plan, do, check, act (PDCA) or "Deming cycle". The PDCA approach is suggested because of its versatility in driving incremental changes for continuous improvement of the systems, processes and operational activities in an enterprise (Owens, 2006). The implementation of the PDCA cycle represents a guideline for the successful implementation of a 5S programme through a more efficient implementation process. The PDCA cycle suggests the following phases: (1) Plan—The programme committee should be organized and a programme plan for each of the 5Ss should be defined; (2) DO—Training and education are provided to every employee. Cleaning and organization of own working area is done as per the schedule; (3) CHECK—The results of 5S are evaluated; (4) ACT—Self-examination is undertaken and corrective and preventive actions taken.

4.1.2 Implementation of the 5S Tool

The 5S has been used by companies to establish and maintain a quality environment in an organization (Khamis et al., 2009). The first application of 5S in a business was in 1980 by Osada (1991). 5S lean tool implementation can fall broadly into one of two categories (Onkar & Bhatia, 2016):

1. *Integrated-with-the-process model*—within this category each department manages the 5S initiatives attending to its specificities.
2. *Stand-alone model*—corporate-wide, the 5S initiatives are implemented and managed by independent teams outside the target department.

The successful implementation of 5S requires its alignment with strategic plans and priorities and optimizing resources to add value. The implementation of 5S could adopt also different models: the *integrated-with-the-process* model or a *stand-alone* model. In the *integrated-with-the-process* model, the 5S elements are embedded within daily operational processes. In the *stand-alone* model, 5S is executed independently of processes. Table 4.2 compares the integrated-with-the-process model and the stand-alone model.

The *integrated-with-the-process model* is more recommended for 5S implementation as it provides timely feedback, visible accountability and the opportunity for immediate mitigation, thus promoting the dynamic interaction of the 5S elements within the system. The 5S *stand-alone model* may require more manoeuvring by team members to ensure the dynamic relationship with the processes.

A guideline for the implementation of 5S is presented next, with a description of the steps required.

4.1.2.1 Step 1: Seiri or Sort

Seiri is associated to a set of practices that are performed in the workplace to make it more functional by identifying or eliminating all unnecessary items from the workplace. These practices are: (1) to discover and identify items which are not needed and not necessary to complete work; (2) develop criteria for the disposal of items not needed; (3) take "before" photographs wherever required; (4) tag the items not needed; (5) find a holding area for red tagged items; (6) classify the items by frequency of use; (7) keep items or equipment used hour by hour or day by day within arms' reach of the point of use; (8) keep items or equipment used once a week or once a month within the work area; (9) store items not needed or unnecessary in the holding area; (10) ensure the holding area is clearly visible and marked to assure visual control of items; (11) display pictures of items and place on a public board visible to all; (12) assign responsibility for the holding area to someone at the beginning of the sorting activity; (13) dispose of all items which are broken or have no value; (14) take "after" photographs wherever required.

Table 4.2 5S implementation models

	Integrated-with-the process model	Stand-alone model
Graphic representation	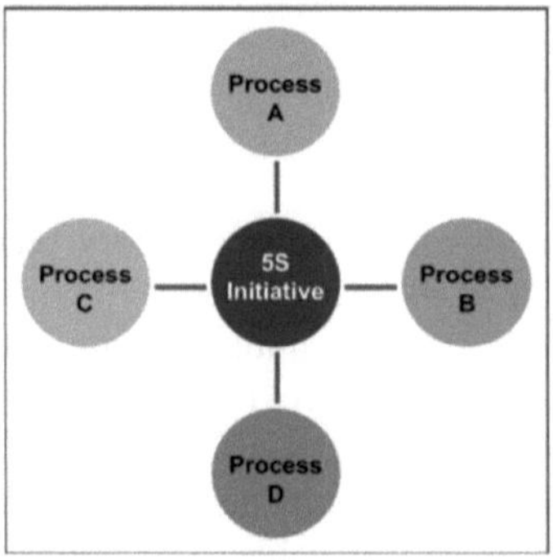	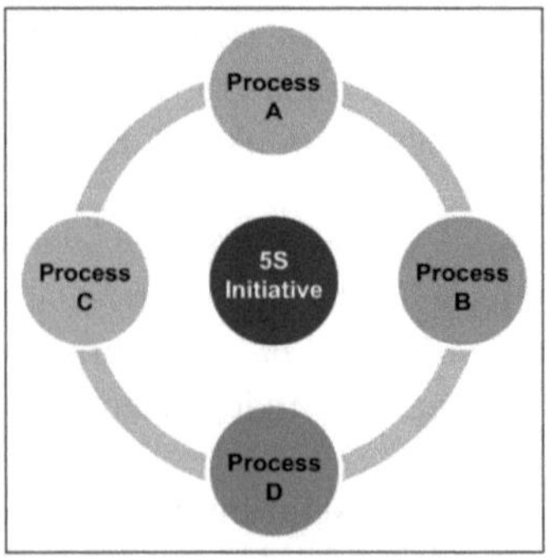
Example	Assembly operators check line clearance (e.g. only acceptable parts needed for the batch in process should be on the assembly line for processes A–D) prior to the start of each batch assembly per operation start-up log.	5S team members (not necessarily the process owners) use a 5S checklist to verify area compliance with sorting non-conforming parts and the organization of each assembly line once a week.
Measure of success	First article inspection or scrap rate trend due to wrong material.	Completion and correction of findings listed in the 5S checklist.
Point of differentiation	In this example, 5S elements (i.e. verification of correct parts) are integrated in the operational execution of processes A–D (i.e. the performance of line clearance). Thus, this is an example of an integrated-with-the-process model.	In this example, 5S elements (i.e. verification of sorting and organization of parts) are performed as independent tasks and not as part of processes A–D. Thus, this is an example of a stand-alone model.

Source: Onkar and Bhatia (2016)

4.1.2.2 Step 2: Seiton or Systematize

Seiton entails putting every necessary item in good order and focuses on efficient and effective storage methods. The main practices associated with this step are: (1) eliminate all unnecessary items from the workplace; (2) decide the place of everything attending to the work flow; (3) taking "before" photographs wherever necessary; (4) using the 5Whys to decide where each item belongs; (5) locating necessary items so they can be retrieved immediately; (6) informing everybody at the workplace about the positioning of the items; (7) making a clear list of items with their locations and put it visible to everybody; (8) outlining the location of equipment, supplies, common areas and safety zones with lines; (9) identifying all necessary items with labels; (10) taking "after" photographs.

In marking the production area, there is a standard range of colours and lines used to identify different zones, such as walkways, storage areas for processing parts, dangerous areas, etc. (Table 4.3).

Table 4.3 Standard for marking production areas

Role	Delineates the walkways	Delineates the storage areas of the processing parts	Delineates the storage areas of finished parts	Delineates the storage areas of rework parts	Delineates the storage areas of rejected parts	Delineates the storage areas of trolleys, carriages, boxes, devices, molds	Delineates the storage areas of waste products	Delineates the dangerous areas	Delineates the fire extinguishers areas
Marking colors									

Source: Onkar and Bhatia (2016)

Table 4.4 5S Check sheet

Machine Number:			Mon	Tue	Wed	Thu	Fri
Machine Name:			3/11	3/12	3/13	3/14	3/15
Machine Location:							
No.	**Checks**	**Frequency**	**Initials of Person Responsible**				
1	Red tag contaminated items	Daily					
2	Remove residue from valves	Daily					
3	Check oil level	Every Tuesday					
4	Apply grease to transfer side	Every Thursday					
Check	Supervised by Bill Smith	Daily					

4.1.2.3 Step 3: Seiso or Sweep

Seiso translates activities associated to cleaning, keeping the workplace clean and inspect the workplace and equipment for defects. To operationalize this step, the following actions should be performed: (1) make a "before" photographs report; (2) adopt cleaning as a daily activity and as a part of inspection; (3) implement practices to avoid dirt and contamination; (4) clean both inside and outside on a daily basis; (5) identify and tag every item that causes contamination; (6) use cause-and-effect methods to find causes of such contamination and take appropriate corrective and preventive action; (7) keep a record of all areas to be improved; (8) maintain 5S check sheets on a daily basis (see example in Table 4.4); (9) create a chart to plan all the activities with the identification of the corresponding responsible; (10) take "after" photographs; (11) assign a periodic 5S time, in addition to the *Seiso* daily' time; (12) undertake a complete evaluation of 5S implementation with the person responsible for all the company or the responsible for the area where this tool was implemented.

4.1.2.4 Step 4: Seiketsu or Standardize

Seiketsu is concerned with the visual control and guidelines for keeping the workplace organized, orderly and clean through the following steps: (1) taking a "before" photographs report ; (2) checking that the previous three steps are implemented properly; (3) creating a 5S board to visualize publicly all team activity checklists; (4) establishing routines and standard practices for systematically repeating the first three Ss; (5) creating procedures for frequently evaluating the status of the first three Ss; (6) standardizing red tag procedures and assigning area rules (see *Seiri*); (7) standardizing activities for creating shadow boards and position lines and labelling all items (see *Seiton*); (8) standardizing cleaning schedules using the "5S Check Sheets" (see *Seiso*); (9) assigning responsibility to individuals for a work area and machinery; (10) ensuring regular inspection and evaluation by a special team; (11) taking "after" photographs and posting them on the 5S board(s); (12) evaluation of 5S levels implementation.

4.1.2.5 Step 5: Shitsuke or Self-discipline

Shitsuke represents a set of activities to ensure the respect for the 5S standards. Some actions related to this last S (self-discipline) are: (1) every employee should treat the workplace as their own home; (2) constant involvement of managers to ensure that the first four Ss are implemented perfectly; (3) employees must make it a part of their daily work and not an action forced upon them; (4) dedication, commitment, devotion and sincerity are mandatory to a successful implementation of 5S; (5) senior management should review periodically the status of 5S; (6) inspections of first three Ss should be performed and the results displayed on the 5S board regularly; (7) single-point lessons should be used to communicate the standards for how 5S work should be done; (8) a root cause problem-solving process should be in place whereby root causes are eliminated and improvement actions are undertaken, including prevention.

When fully implemented, the 5S lean tool has a positive impact on organizations, such as increasing the morale of workers, creating a very positive image of company to the customers and increasing its efficiency. Not only will work conditions be improved, but also the company's operations will be performed better as 5S has a positive effect in terms of continuous improvement, reducing waste and increasing quality and decreasing lead times. 5S is not only a system for housekeeping or a list of action items, but is an integrated concept of actions, conditions and culture. A successful implementation of 5S, requires that all employees have a deep understanding of each "S".

4.1.3 Evaluation of the Success of a 5S Project

Table 4.5 provides an overview of the 5S tool, with a brief characterization of each step, main benefits that companies can attain with them and problems

Besides the five steps associated with the 5S methodology, normally it is implemented firstly by defining a cross-functional team, then touring all areas associated with the focused manufacturing process and finally brainstorming ways to improve the organization to reduce waste.

Table 4.5 Overview of 5S

S	Characterization	Advantages	Problems identified before implementation
Sort	Remove unneeded items from production operations.	Better management of space, time, money, energy. Minimization and mitigation of problems. Improving communication flows. Increasing product quality. Enhancing productivity.	Factory overloaded and hard to work in. Storage of unnecessary items. Time wasted searching for parts/tools. Negative impact of excess of inventory and machinery on costs. Excess stock hides production problems. Excess of items and equipment difficult the process flow.
Set in order	Arrange and label needed items so that they are easy to find and use.	Eliminate many kinds of wastes associated with searching, difficulty in using and returning items.	Excess of motion. Workers are always searching for something. High levels of inventory. High level of defective products. Hazardous working conditions.
Shine	Keep everything swept and clean every day.	Turn the workplace in a place in which everyone will enjoy working.	Lack of sunlight in the workplace. Oil and water in the middle of the workplace Lack of machine and equipment maintenance.
Standardize	Integrates Sort, Set in Order and Shine into a unified whole.	Ensures that conditions do not deteriorate and facilitates implementation of the first three pillars.	Workplace returns to its previous undesirable conditions with work areas dirty and cluttered.
Sustain	Making a habit of maintaining correct procedures.	Positive consequences of keeping the procedures	Piling up of unnecessary items. Tools and jigs are put all over the places becoming the workplace unsafe. Dark, dirty, disorganized workplace.

A variety of methods have been suggested for the evaluation of projects: cost benefit analysis (CBA), which became the accepted standard, is the most widely used (Odgaard, Kelly, & Laird, 2005).

With particular regard to projects and programmes, evaluation is the assessment and the analysis of the effectiveness of an activity; it involves the formulation of judgments about impact and progress. Evaluation is the comparison of the actual effects of a project against the agreed planned ones. Evaluation looks at what was

Table 4.6 5S project evaluation and sustainment

Week of ________ Evaluator:______________ Location:______________

For each "S", circle the block that best describes the location being evaluated.

	Points					
	5 Best-in-Class	4 Good	3 Fair	2 Marginal	1 Poor	0 Unacceptable
Sort	Every item in the area is needed.	Minimal clutter, small amount of unneeded items.	Some clutter. Some items stored on top of cabinets. Some needed items are not present.	Clutter and unnecessary items on desks, shelves and/ or cabinets.	Generally cluttered. Many unnecessary items on desks, shelves, cabinets and/or floor.	Station is cluttered/unsafe. Majority of unnecessary stations are cluttered/unsafe. Abundance of unnecessary items, including furniture and equipment.
Set in order	Everything in the area is in great condition and operates perfectly. No dust or dirt, clean floor.	Everything in area is clean and functional. Floors clean. Recycle bins used and emptied daily.	Most items in area are clean and functional. Some shared equipment doesn't work. Floors are moderately clean.	There is some clutter. Some items are not clean (floors, waste containers).	Area needs to be picked up. Many items need to be cleaned (floors, waste containers).	Area is dirty. Possible health and safety hazards exist.
Shine	Everything in the area is in great condition and operates perfectly. No dust or dirt, clean floor.	Everything in area is clean and functional. Floors clean. Recycle bins used and emptied daily	Most items in area are clean and functional. Some shared equipment doesn't work. Floors are moderately clean.	There is some clutter. Some items are not clean (floors, waste containers).	Area needs to be picked up. Many items need to be cleaned (floors, waste containers).	Area is dirty. Possible health and safety hazards exist
Standardize	5S checklists for all personal/shared areas are available and followed. Opportunities for improvement noted – all resolved. Metrics and charts in area are current.	5S checklists for most personal/shared areas are available and followed. Opportunities for improvement are noted – most resolved. Metrics/charts are current.	5S checklists for most personal/shared areas are available and followed. Opportunities for improvement are noted – some resolved. Metrics/charts in the area are current.	5S checklists for personal/shared areas are not available or not followed. Assessment done weekly, but no opportunities for improvements noted. Metrics/charts in the area are not current.	Very few 5S checklists exist for the area. 5S assessment not done week prior. Unclear if actions taken to resolve problems. Metrics/charts not current.	No 5S checklists exist. 5S assessment not done within previous week. No evidence of follow-up actions to problems. Metrics/charts are not current.
Sustain	Clearly documented and followed system to maintain 5S. World class performance sustained for at least 6 months.	System in place and strong interest in maintaining 5S. Good performance sustained for at least 3 months.	System in area and moderate interest in maintaining 5S. Minimum performance sustained for at least 1 month.	Little system in area, little interest in maintaining 5S. Minimum performance sustained for at less than 1 month.	No system in area, little interest maintaining 5S. Very little stability in new 5S performance.	No system or commitment within organization to sustain 5S

# of 5s circled: _____ *X5=*______
# of 4s circled: _____ *X4=*______
# of 3s circled: _____ *X3=*______ **5S RATING:** ____
# of 2s circled: _____ *X2=*______
# of 1s circled: _____
TOTAL: ______

Opportunities for improvement: who... when
1
2
3

Source: CDOT (2017)

planned, what has been achieved and how it has been achieved (Archibald, 2012; Shapiro, 1996).

For the evaluation of a 5S project some tools can be used to monitor the state of the art of the implementation level for each S. For example, the CDOT office of transportation Safety and Continuous Improvement Program (CDOT, 2017) suggests a tool that works as a checklist (Table 4.6).

The evaluation of projects can be implemented throughout the whole duration, attending to three stages: ex-ante, interim and ex-post. The first stage aims to compare, select and finance alternative projects. The second aims to improve strategy or processes. The third aims to adopt lessons, insights, judgment and awareness regarding the decisions taken and implementation of projects (Linzalone & Schiuma, 2015).

4.1.4 5S Practical Application

This section illustrates the implementation of the 5S tool in an automotive company which makes up part of an automotive group producing the full line of components that go into automotive seating. At the beginning of the implementation process of 5S, a team was created. This team had as its main objective working on the shop floor environment and on the optimization of production.

First, the working team participated in a 5S workshop to understand how the tool works and identify the area for intervention and the objectives to be achieved. In this first stage, the choice was made to act in a production line, later designated as a "model line". Thus, the implementation of the 5S tool was undertaken in a single production line and only after validation was it implemented in other production lines.

The 5S implementation team was created by the project manager, who defined the facilitator and the pilot member. The role of the facilitator is to provide support to the 5S implementation group, supporting the members in daily inspections and audit tasks, as well as interacting with management to obtain support and resources for the efficient application of this methodology. The pilot member is responsible for creating the team, identifying the knowledge gaps in the 5S team, defining the priorities and resources required to carry out its implementation and maintaining deadlines.

The person with responsibility for administering the application of 5S should define the area in which it will be implemented, the deadline and the objectives to be attained. All this information is placed in the panel for the task group defined (Fig. 4.1). This panel must be placed in the application zone, visible to all employees of the organization, as defined during the first meeting of the committee. Thus, whenever an improvement activity is initiated, it is used as a communication panel.

This teamwork aims to audit the intervention zone. This first step is strategic for the identification of problems that may exist in that zone, such as dirty walls, issues with oil-pouring machines, loose wires, damaged structures, tools and machines that are not used, among others.

After the analysis of the zone in which the intervention will be undertaken, the next step is the formalization of the actions through the following steps:

1. Definition of the periodicity of meetings—a 1-h weekly meeting was established.
2. Definition of the list of priorities, with dates and places for intervention, starting with the most urgent actions (productivity, quality, safety...), then moving on to

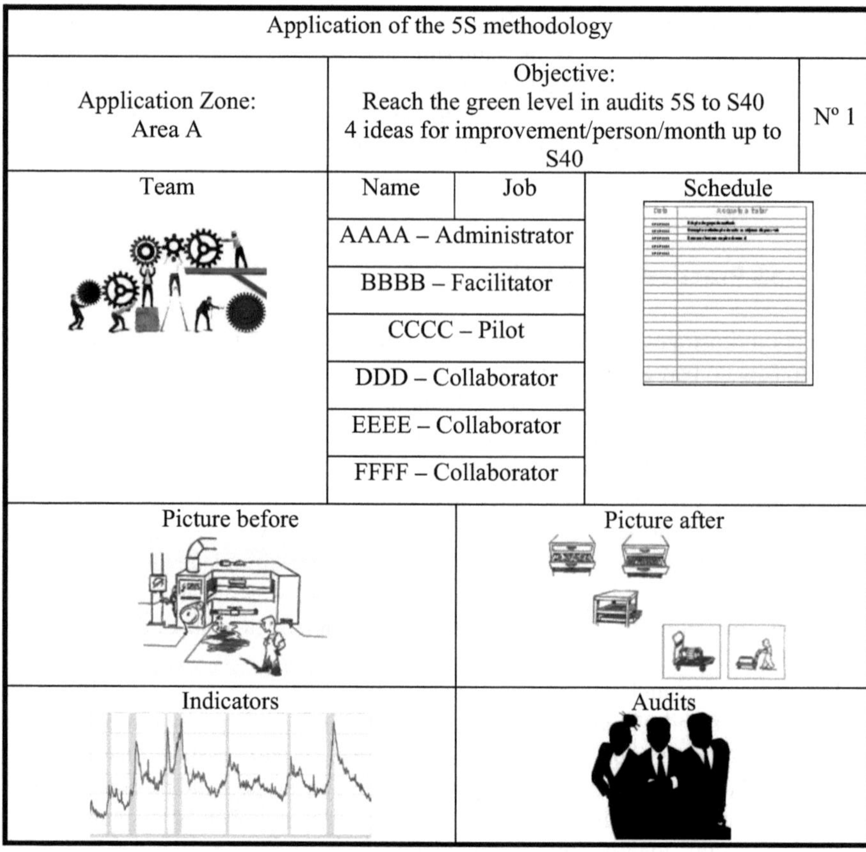

Fig. 4.1 Task panel group

the less urgent ones. After the agreement of all the stakeholders, the timetable is drawn up and disseminated.

3. Preparation of the 5S audit ranges, with the identification of the points to be audited and how to follow up on their evolution.
4. Definition of standards.

4.1.4.1 Evolution of the 5S Implementation

Each supervisor audits a different area every week to ensure awareness of the reality of the different production zones inside the plant. The audit checklist is divided into five parts, each corresponding to an "S"; it is only possible to progress to the next "S" if the previous one is reached at 100% (Table 4.7).

The follow-up of the evolution of the 5S implementation was performed on a weekly basis by each production supervisor. The document used to follow up on the evolution of the 5S implementation in the production zone is illustrated in Fig. 4.2.

Table 4.7 5S audits

			5S AUDITS	Responsible for the area: Area: Auditor:			
				Yes	No	N/A	Observation
1° S	Sort	1.1	There is no useless or outdated object on the workstation.				
		1.2	Does not exist (tools, components, documents,) outside your site				
		1.3	The corridors are free and without security risks				
		1.4	There are no sharp or sharp objects in people's movement areas.				
			Result: (∑ Yes / Total number of questions – No) x 100= _______	Note: To go to the next step you have to get the result of 100%			
2° S	Set in order	2.1	Is there a zone and is it defined?				
		2.2	Is everything in the work zone identified and tidy?				
		2.3	Are electric wires and air hoses insulated and grouped not impeding movement?				
		2.4	Are extinguishers and safety zones identified, visible and accessible?				
			Result: (∑ Yes / Total number of questions – No) x 100= _______	Note: To go to the next step you have to get the result of 100%			
3ª S	Shine	3.1	Is the work station clean?				
		3.2	Are the limits of the zones in good condition and visible?				
		3.3	Is all the garbage separated and placed in identified locations for each situation?				
		3.4	Are the measuring instruments (manometer, oil level ..) kept clean and functional?				
			Result: (∑ Yes / Total number of questions – No) x 100= _______	Note: To go to the next step you have to get the result of 100%			
4° S	Standardize	4.1	Is there a standard for each work zone?				
		4.2	Are the schedules respected?				
		4.3	Are the definition of production levels and follow-up respecting the standards?				
		4.4	Are the documents used standardized?				
			Result: (∑ Yes / Total number of questions – No) x 100= _______	Note: To go to the next step you have to get the result of 100%			
5° S	Sustain	5.1	Are the limits of the production area respected?				
		5.2	Is the production follow-up board up to date?				
		5.3	Is the rapid response board up to date?				
		5.4	Is the 5S working group board up to date?				
		5.5	Is there a daily follow-up of the improvement ideas made by supervisors?				
			Result: (∑ Yes / Total number of questions – No) x 100= _______	Note: To go to the next step you have to get the result of 100%			
			Total Result: (∑ Yes / Total number of questions – No) x 100 = _______				

As can be seen from Fig. 4.3, all the zones that make up part of the layout of the plant were identified. This plant has eight lines of production, two zones with presses, one zone for maintenance, one logistics zone (performing logistics operations associated with warehousing and reception) and some production offices. Also, different colours were used to identify these different zones, attending to the type of "S" implemented in each zone (1S—*seiri*, 2S—*seiton*, 3S—*seiso*, 4S—*seiketsu* and 5S—*shitsuke*).

A worksheet was implemented for all machines, equipment, lines and production environments (Fig. 4.3). This worksheet makes it possible to identify the intervention zone (presses), the focused operation (cleaning) and an illustration of the equipment with the identification of the tasks, the responsible person, frequency and the way in which the different tasks are performed. This provides important support for the implementation of 5S.

The evolution of the results based on the implementation of 5S and ideas for improvement were also registered graphically, as shown in Fig. 4.4.

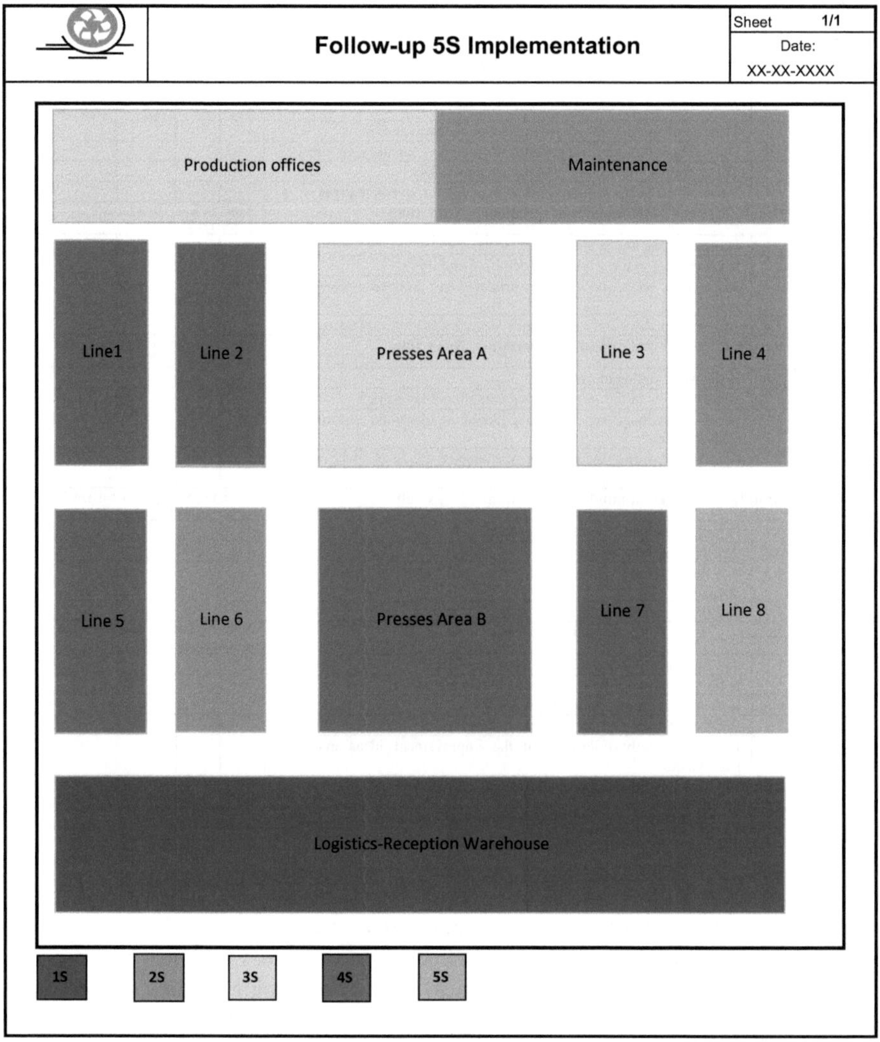

Fig. 4.2 Follow-up of 5S evolution in the production zone

As can be seen from Fig. 4.5, the implementation of 5S increased over the 40 weeks of intervention. The goal of 80% for the implementation level of 5S was only reached in the 37th week, this target being exceeded in the next 3 weeks. Moreover, the number of ideas for improvement is very irregular; however, by the end of this period they had increased and become more stable. Thus, the 5S methodology represents a very useful tool in the implementation of a JIT factory as it allows the proliferation of rigorous thinking among all employees, providing organizational means that lead to high levels of quality, safety, productivity and employee involvement.

	Gama 5S	Sheet	1 / 1
		Date	
		XX - XX -XXXX	
GAP:	Application area	Press ATP	
Presses	Operation	Cleaning	

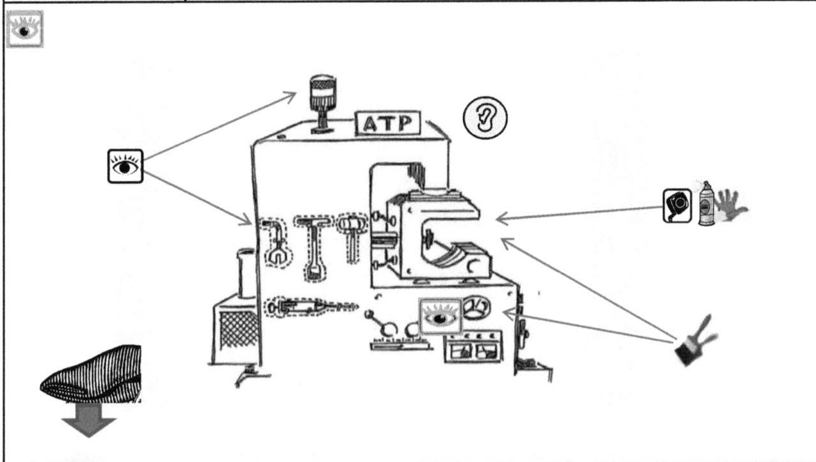

N°	Designation	Person	Frequency	Means
1	Check that all tools are in the indicated location	Operator	1 time/shift	
2	Cleaning of molds	Operator	1 time/shift	
3	Detection of strange noises	Operator	Whenever the machine is used	
4	Machine painted white	Operator	Quarterly	
5	The floor must be clean and free of dirt	Operator	1 time/shift	
6	The work area must be free of dust	Operator	Whenever necessary	
7	Verify that the identifications and documentation are up to date and in good condition	Operator	End of turn	
8	Check gauge identification (within limits)	Gap Leader	Whenever necessary	
9	Check if all rules have been met	Gap Leader	End of turn	

Validation

Supervisor	Quality	Gap Leader
Date:	Data:	Date:

Changes:

1: XX-XX-XXXX

Fig. 4.3 5S process sheet

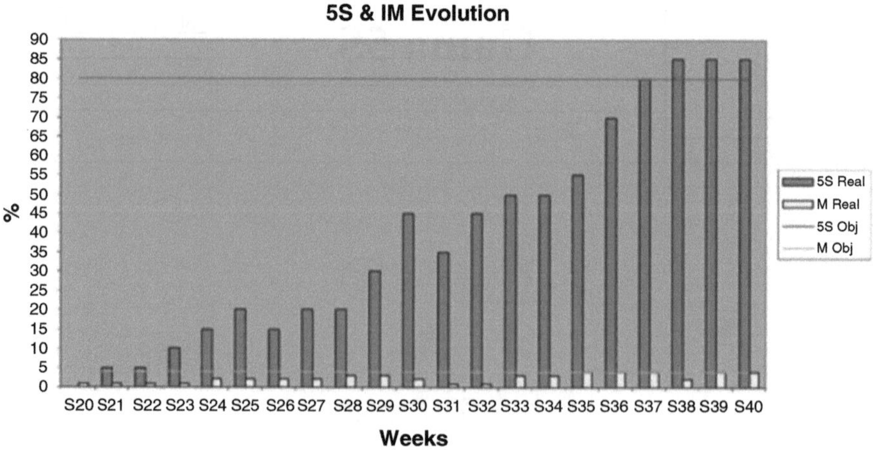

Fig. 4.4 Evolution of 5S implementation and ideas for improvement

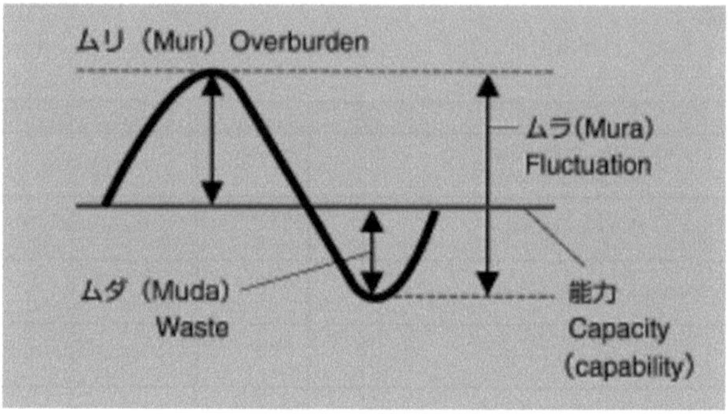

Fig. 4.5 Muda, mura and muri and their interrelationships. Source: Narusawa and Shook (2009)

4.2 Standardized Work

Imai (1986) in his seminal work said that there can be no improvement without standards. Moreover, he considered that standardization is actually the starting point for any improvement. Standardized work is very important in a continuous improvement culture because it allows a basis for evaluation that would not be possible if the task was performed differently each time. Also, the establishment of standardized processes and procedures is considered by many authors to be key to creating consistent performance (see e.g. Liker & Meier, 2006). Thus, it is only when the process is stable that the creative progression of continuous improvement can begin. Toyota considers the development of standardization to be a baseline for continuous improvement, which means that future results should improve the standard (Liker & Meier, 2006). When a standard is improved, this new standard turns into the baseline for future improvements and so on. Standardized work is also the basis for stability, facilitating the adoption of JIT and jidoka.

Work standardization can be defined as the currently best-known method for accomplishing work (Martin & Bell, 2011) by creating specific procedures for each operator's work in a production process (LEI, 2008). Misiurek (2016) incorporates new insights into the definition of work standardization, defining it as a process for describing the best current way of performing a given operation, improving on this and training operators in it; this is supported by three key elements: process documentation, process improvement and operator training. It includes writing down the best currently known way of performing operations in any workplace with consideration for occupational health and safety, quality, ergonomics and correctness, then training all employees.

According to Martin and Bell (2011), work standardization is not just a format for the documentation of work. Rather, the authors see it as a basic TPS philosophy that is intertwined with kaizen. Indeed, the standardization of work is a vital element in lean manufacturing and the attempt to eliminate waste, unevenness and unreasonableness (Ohno, 1988a, 1988b, 1988c). The Toyota Motor Corporation (1998) defines the standardization of work in the following way: "Standardized work is a tool for maintaining productivity, quality, and safety at high levels. It offers a reliable framework for executing work at the required takt time and for identifying improvement opportunities in work procedures. Moreover, it establishes detailed, step-by-step guidelines for every job". As standardized work encompasses following procedures consistently, any intrinsic problems in the working sequence occur repeatedly and are clearly identified. Therefore, team leaders and operators can easily identify problems and solve them promptly. At Toyota, standardized work means, in essence, that a process is actually operating as specified by the standard, where a standard is a description of how a process should operate. The standard is the pre-specified, intended and normal pattern (Rother, 2010).

Standardized work, which should be done with operators and for operators, is a powerful vehicle that contributes to fostering and disseminating knowledge. Moreover, according to Patchong (2013) the deployment of standardized work is the DNA coding of operational excellence—process by process, workstation by workstation, sequence by sequence—wherever there is a human being. In practice, standardized work is typically associated with discipline, repetitiveness, predictability, consistency and reproducibility. Ohno (1988a, 1988b, 1988c) claims that the elements to be consider in standardized work are workers, machines and materials. Moreover, he argues that if not combined effectively, the workers will feel alienated and find it impossible to produce efficiently.

There are many benefits that can be achieved with standardized work if it is used properly. Some important benefits are: (1) the creation of a reference point from which to improve continuously; (2) clear start and stop points for each process; (3) process control; (4) variability reduction; (5) the generation of insights and a rhythm for work to prevent overburdening; (6) improved quality and flexibility; (7) process stability (i.e. predictable outcomes and repeatability); (8) transparency in abnormalities; (9) clear expectations; (10) audit and problem solving; (11) documentation of the current process fostering preservation of know-how and expertise; (12) helping people structure their work; (13) easier training of new operators; (14) reductions of injuries and strain; (15) a platform for individual and organizational learning (Dennis, 2015; Emiliani, 2008; LEI, 2018). Indeed, as this tool is closely connected to the elimination of root causes, it effectively affects the reduction of human errors (Misiurek, 2016). Moreover, standardized work ensures safe working practices and the training of people and acts as a basis for addressing *muda* (waste), and *muri* (overburdening) and the identification of *mura* (unevenness/fluctuation) (Fig. 4.5).

It is generally accepted that although standardized work is fundamental to the stability of processes and performance, this is probably one of the most difficult tools to sustain. One of the reasons for this is the impossibility of maintaining a level of process performance. According to Rother (2010), a process will naturally tend to deteriorate, even though a standard is created and is explained to everyone involved and posted. This is due to interaction effects and entropy that dictate that any organized process naturally tends to decline to a chaotic state if it is left alone. Thus, a key aspect of standardized work is training and auditing. Standardized work is mainly the process of raising people's awareness through training and making work improvements (Misiurek, 2016).

The origins of standardized work can be traced back to the Training Within Industry (TWI) service, this being an essential part of the TPS (Graupp & Wrona, 2010). The TWI service was an emergency service created by the US Government War Production Board aimed at helping industry become more productive than had ever been thought possible. The TWI programme was developed to support defence industries in satisfying their manpower needs through worker training so as to reap the maximum benefits from using their best skills up to the maximum of individual ability (Graupp & Wrona, 2010). At that time, many skilled workers were on active duty and it was necessary to develop an effective training procedure to train quickly and efficiently unskilled people to perform their work (Liker & Meier, 2006). This programme sought to standardize the way people were formed in a new task.

One key building block for Toyota from 1951 to 1953 was the adoption of the three main TWI courses of Job Instruction (JI) (War Production Board, 1944a), Job Relations (JR) (War Production Board, 1944b) and Job Methods (JM) (War Production Board, 1943). This was the first structured management and supervisory training programme rolled out through the company worldwide in Toyota. Moreover, the basic concept of TWI, especially related to JI, underwent decades of continual refinement and application within Toyota (Graupp & Robert, 2016).

TWI was focused on supervisor training, preparing them to train others. A supervisor is anyone who is in charge of people or who directs the work of others, including, as an example, managers, group leaders, team leaders, or even operators who have been assigned to train other operators. The JI element was designed to teach supervisors how they should train employees so that they can easily remember how to perform a job properly, safely and consciously. The JM aspect teaches supervisors ways of improving job methods so as to yield larger quantities of quality products, more rapidly, through the best use of the currently available manpower, machines and materials. Finally, the JR element prepares supervisors to lead people and provides them with an analytical method to solve problems effectively (Graupp & Robert, 2016). Thus, JI promotes the development of standards and their spread throughout the organization, while JM supports improvement in standards and JR boosts process innovation and problem solving.

The training method used by Toyota today is basically an imitation of the material created in the US in the 1940s. Toyota has made only a few minor additions and now uses the material to train effectively thousands of workers. Moreover, all leaders within Toyota are required to learn the JI training method (Liker & Meier, 2006).

4.2.1 Implementation of Standardized Work

Standards should not be forced down from above, but rather established by production workers themselves (Ohno, 1988a, 1988b, 1988c). It is generally accepted that standardization should be centred on tasks that involve human motion, are repeated frequently or have been selected for standardization, for example due to high complexity and so on. Indeed, most applications of standardized work have been devoted to high volume and moderately standardized items. Under such conditions, tasks become highly standardized, stating to the second what the operator must do (Lander & Liker, 2007). Thus, the application of standardized work is particularly relevant in the case of repetitive tasks and those involving movement. Nonetheless, it can also be applied to the standardization of noncyclic tasks (Martin & Bell, 2011). As a rule of thumb, standardization should be focused on the 20% of processes that represent 80% of variability.

According to Dennis (2015), it is impossible to work to standards when there are continuous line stoppages and slowdowns. Thus, prior to standardizing a process, it is essential to work on its stability. Some sources of variability are quality problems with incoming parts, problems with machinery, jigs, or tools, parts shortages and safety problems (Dennis, 2015).

The implementation of standardized work consists of maintaining, at a high level, the productivity, quality and security of a particular operation. During the initial application of standardized work, the aim should be to create a baseline for improvement. Standardization is not only applicable in manufacturing operations, but is equally applicable to other types of operations (e.g. in setup operations), functional areas (such as purchasing and supplier development, engineering or new product development) or systems (e.g. services). Moreover, it can be applied at different levels of decision making: the operational level (e.g. in the standardization of the operators' workstation work), tactical level (e.g. in the standardization of the team leader's work), or strategic level (e.g. in the standardization of the work of executive leadership). However, the challenges in the implementation of standardized work increase from the operational to the strategic level.

Liker and Meier (2006) identify three types of prerequisites for the application of standardized work: one from the point of view of the task, which must be repeatable (meaning that the same result is achieved each time it is executed); the second from the point of view of the resources/equipment, which must be reliable and have the

Table 4.8 Process flow chart form

Process Flow Chart											
Process analysis											
		Process activities					Time	Waste/value			
N.°	Process steps	Operation	Transport	Inspection	Delay	Storage		Muda	Muda (necessary)	Value added	Description
1		○	⇨	□	D	▽					
2		○	⇨	□	D	▽					
3		○	⇨	□	D	▽					
4		○	⇨	□	D	▽					
5		○	⇨	□	D	▽					
6		○	⇨	□	D	▽					
7		○	⇨	□	D	▽					
8		○	⇨	□	D	▽					
9		○	⇨	□	D	▽					
10		○	⇨	□	D	▽					
11		○	⇨	□	D	▽					
12		○	⇨	□	D	▽					
13		○	⇨	□	D	▽					

minimum downtime; the third and last, from the point of view of quality, which must have minimal defects and be consistent in terms of the key parameters.

Documents are very important for implementing standardized work, but by utilizing only one document the results will be limited, as no document deals with all the elements of the approach; thus, it is necessary to employ combined application to achieve meaningful results (Marksberry, Rammohan, & Vu, 2011). In subsection 4.2.3 the set of core documents used in standardized work is presented.

Patchong (2013) proposed the following four steps for the deployment of standardized work: (1) capturing the current state; (2) improving the process; (3) training; (4) application and auditing. Step 1 is supported by a process study and analysis and by standardized work support documents. Process study and analysis is used to define the actual time required for each individual work element and to estimate existing capacity and identify the main causes of capacity losses. The first step in a process study is work observation and the division of the work into a set of work elements required to produce one item. Afterwards, each work element is analysed to identify and eliminate waste. In the final step, a time study is performed to estimate the actual time required for each individual work element. Two valuable forms for supporting a process study are the process flow chart (Table 4.8) and the process study sheet (Table 4.9). Another valuable tool for a process analysis is the process capacity sheet, which is used to calculate the capacity of each machine in the process to confirm true capacity and to identify and eliminate bottlenecks. This sheet is presented in Sect. 4.2.3.

Table 4.9 Process study sheet

Process Study	Process:						Product:			Observer:			Date/Time:	Page /
Process Steps	Work Element	OPERATOR Observed Times											MACHINE Cycle Time	Notes
		1	2	3	4	5	6	7	8	9	10	Repeatable		

Source: Narusawa and Shook (2009)

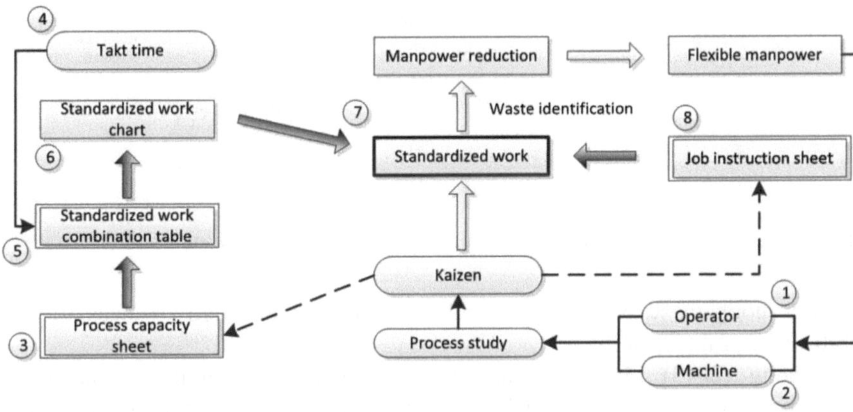

Fig. 4.6 Standardized work creation steps. Adapted from: Narusawa and Shook (2009)

As previously mentioned, prior to the standardization of a process, a certain level of process stability is important in order to reap the benefits of standardization. Thus, during step 1, grounded on the process study and analysis, process stability should be addressed.

When the process is sufficiently stable, the standardized work support documents are created, namely the standardized work chart, the standardized work combination table and the operator work instruction sheet. In Fig. 4.6 the approach to be followed during the creation of standardized work is presented.

In step 2 the focus is placed on rapidly improvements based on shop floor observations. The major types of improvements are centred on operator work methods, operator work conditions and equipment availability. Figure 4.7 suggests a procedure for the identification of improvement opportunities and assurance of standardized work. It should be noted that the systematic observation of the operator working cycle is the basis for improvement, leading to the identification of causes of

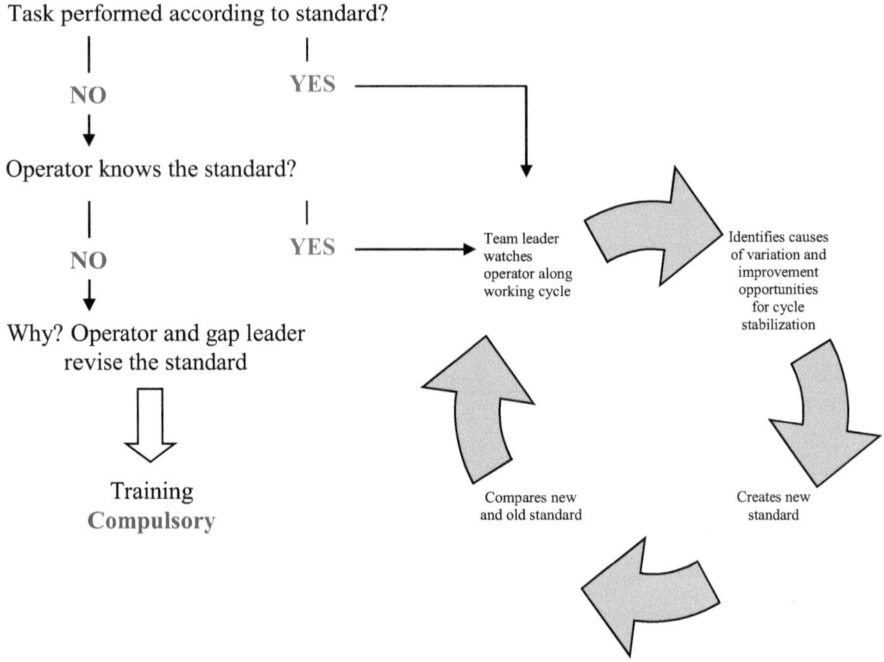

Fig. 4.7 Standardized work assurance and improvement opportunities identification

variation along the cycle time, as well as to the identification of ideas for the stabilization of the cycle.

Patchong (2013) highlights that it is not enough to improve and write standardized work documents. It is also very important to train operators to use these new methods. Thus, one important issue is the choice of the right training method. In addition to this, the same author argues that to perpetuate continuous improvement of existing work methods, auditing is the only way to identify problems and resolve them. This is approached in step 4, during which frequent and simple auditing mechanisms should be put into practice.

An alternative to Patchong's (2013) approach in the deployment of standardized work is to apply steps 1 and 2 in an integrated way. In this case, the improvement phase is approached during the process study and analysis and during the creation of standardized work support documents, looking for an optimized way to perform the work elements.

In general, Toyota has approached improvements through the OSKKK methodology, which establishes an order for the improvement activities accomplishment. OSKKK stands for: observation, standardization, kaizen 1, kaizen 2, and kaizen 3. These are the kaizen steps in the order of effectiveness and sorted from least to most costly to implement.

The observation phase is concerned with the study and analysis of process, aiming to identify the main problems that arise in the given process. During the standardization phase, materials, motions/movements, tasks and management are standardized by prioritizing the critical parts of the process. Kaizen 1 is focused on the improvement of information, material flow and process, using value stream mapping. Kaizen 2 is dedicated to equipment improvement (such as reducing changeover time and accelerating machine speed). Finally, kaizen 3 is connected to the improvement of layout.

Ideally, standardized work should be created for at least three scenarios (scenario n, scenario n−1 and scenario n+1). This allows the adjustment of standardized work to different takt times. Basically, scenario n is devised considering the average customer demand. Scenario n−1 is established for periods of smaller demand (minimal demand), leading to a higher takt time and possibly to production with a smaller number of operators. In contrast, scenario n+1 should be used during periods of high demand (maximal demand), in which the takt time is smaller and the number of operators may be increased.

After the creation of a new standard, or after adjustment of a current standard, all the standardized work documents should be placed within the work area, so that in the event of any doubt or for auditing purposes they are easily accessed.

4.2.2 Main Elements of Standardized Work

There are three required components to standardized work (LEI, 2008; Martin & Bell, 2011; Monden, 1998a, 1998b, 1998c; Narusawa & Shook, 2009): takt time, work sequence, and standard in-process stock (Fig. 4.8).

Takt time measures the pace of customer demand, thus representing the desired time that it should take to obtain one unit of production output. Takt time can also be seen as the rate of production that can meet customer demand on time, or the rhythm at which an organization should operate.

German and Japanese axis powers brought this concept to the Mitsubishi aircraft plants in Japan. From there, it progressed to Toyota and thence lean manufacturing. The average time between parts is named after the German word "takt", meaning pulse, stroke, timing, beat, or cycle (Roser, 2017).

Takt time is a customer-driven measure associated with market requirements, representing essentially the maximum time available to satisfy those same requirements. Moreover, it is a reference point for designing work (Liker & Meier, 2006). Takt time is equal to the division of the time available to produce parts within a specific time interval by the number of parts demanded in the same time interval. Equation (4.1) presents the takt time calculation formula and an example of its application is provided in Example 1. The operating time represents the available time per unit of time, for example a total working day time, corrected by the planned stops. Examples of planned stops are paid breaks, lunch time, cleaning time, meetings, planned maintenance and other administrative non-working time.

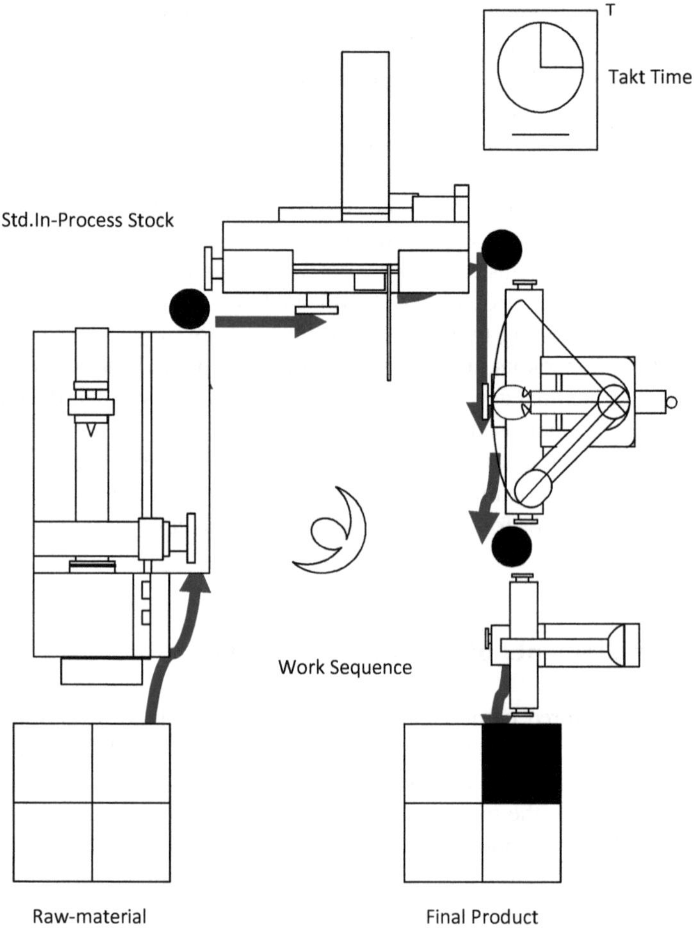

Fig. 4.8 Basic elements of standardized work

$$TT = \frac{Operating\ Time - Planned\ stops}{Customer\ Demand} \tag{4.1}$$

where TT denotes takt time.

Example 1 Takt time calculation
Daily operating time:

- 8 h and 30 min (1 shift)

Planned stops:

- Lunch: 30 min
- Paid break: 10 min
- Cleaning: 10 min
- Start of shift meeting: 5 min
- Planned maintenance: 20 min
- Changeovers: 30 min

Customer demand: 1000 parts per day

According to Eq. (4.1), the takt time will be equal to:

$$TT = \frac{8:30 - (30 + 10 + 10 + 5 + 20 + 30)}{1000} = 24.3\backslash \text{seconds}$$

The takt time value shows that the customer will require a product every 24.3 s. This implies that it is necessary to balance the working system (e.g. the cells or the production lines) to work at this rate. Indeed, the goal of any organization is to work at the customer's pace. For this to be possible, a reliable, problem-free production system is required.

Nevertheless, in practice every organization faces several problems, which may occur in a variety of ways, such as difficulty in supplying production lines on time, lack of reliable machines, constant maintenance problems, absenteeism, etc. If, for some reason, during a certain period the working system cannot comply with the customer's pace, two solutions, among others, may be required to compensate losses and fulfil the customer's requirements in the allotted time. One of these solutions is to make use of overtime. Overtime is perhaps the hardest solution to apply, usually due to labour issues, but it is one of the most efficient. Working extra time means working with a shorter operating time than the customer. That is, instead of operating, for example, over a shift that lasts 8 h and 30 min, during extra time the working system operates for 7 h and 30 min. Thus, considering the same data as in Example 1, the takt time would become equal to:

$$TT = \frac{7:30 - (30 + 10 + 10 + 5 + 20 + 30)}{1000} = 20.7\backslash \text{seconds}$$

This implies working at a higher pace than the customer, but it also leads to working 1 h less. This time results in a safety period preventing the occurrence of any unforeseen events as it is possible to use this hour to comply with the customer's requirements. The other solution is to produce at the pace of the target cycle time (TCT). In this case, as in the previous one, the production rhythm is faster than the takt time to absorb the possible losses. The TCT results from the product of the takt time employing a safety coefficient (SC) that is typically expressed as a percentage of the takt time. The determination of an accurate estimate for its value is a complex task as it requires accurate knowledge of the productive process. To determine this, an in-depth study of what occurs on a daily basis in the working system is necessary as it is imperative to recognize the main problems occurring, as well as their implications for non-productive time. Considering again the data from Example 1 and a safety coefficient equal to 15%, the TCT would be equal to:

$$TCT = TT \times (1 - SC) = 24.3 \times (1 - 0.15) = 20.6\backslash\text{seconds}$$

Clearly, the objective of any organization should be to equal the takt time and the TCT and minimize the overtime.

The takt time concept is strongly connected to the cycle time concept, although takt time is different from cycle time, which is equal to the actual time needed to produce one unit of output. If the cycle time in a given operation is greater than the takt time, a bottleneck will occur and additional time will be needed to meet the customer demand on time. In contrast, if the cycle time is lower than the takt time, there will be overproduction or wait time. Ideally, the cycle time should be roughly equal to or slightly lower than the standard or desired cycle time (Martin & Bell, 2011) and the standard cycle time should equally be roughly equal to or slightly less than the takt time, so that demand is met (Liker & Meier, 2006). A truly lean operation will synchronize takt and cycle times to meet customer demand.

The second basic element of standardized work is the work sequence. This is the sequence in which the operator should perform a set of operations (converting raw materials into finished products), leading to process completion within the required takt time.

The third and last element of standardized work is standard in-process stock, which should be understood as the standard inventory required to keep a process (work sequence) running smoothly. It is defined as the maximum material waiting or flowing through the system when the process is running. Thus, it is the minimum quantity of work-in-process required, within the several workstations and in the machines, so that the process runs without breaks. Thus, the purpose of standard in-process stock is to protect the system against a given problem at a given workstation. However, this stock must be kept at a controlled level, i.e. at the strictly minimum quantity necessary.

4.2.3 Standardized Work Documents

According to the literature (e.g. Dennis, 2015; LEI, 2008; Liker & Meier, 2006), there are three master documents commonly used for creating standardized work: the

Table 4.10 Example of a process capacity sheet

PROCESS CAPACITY SHEET										Date:		
Part Name:		XXXXXXXXXXXXXXXXX								TL	GL	AM
Part #:		XXXXXXXXXXXXXXXXX										
No	Process name	Machine #	Basic Time			Tool Change		CAPACITY (50)	OBSERVATIONS			
			MANUAL (secs)	AUTO (secs)	TOTAL (secs)	Change	SET UP (secs)					
1	Robotic Welding	OP20/30	30	41	71	500	5	50				
2	Robotic Welding	OP40	10	22	32	200	12	112				
3	Robotic Welding	OP50	14	35	49			73				
4	Robotic Welding	OP60	17	31	48			75				
5	Robotic Welding	OP70	15	40	55	200	5	65				
6	Robotic Welding	OP80	8	33	41			87				
7	Screwing	OP100	10	10	20	50	15	177				
8	Leakage test + Packing	OP110	15	20	35	50	20	101				
		TOTAL	119									

process capacity sheet, the standardized work combination table and the standardized work chart. These are used by engineers and front-line supervisors to design processes and by operators to make improvements in their own jobs (LEI, 2008). Whenever this happens, engineers and supervisors must adjust the standardized work in order to meet quality, safety and productivity standards.

The standardized work documents presented above are typically used in conjunction with the job instruction sheet and with other support documents (such as the operator balance chart—also called *yamazumi chart)*. Each of these documents is explained below and an example is also presented.

4.2.3.1 Process Capacity Sheet

The process capacity sheet is used to determine the capacity of each machine in a set of processes to confirm true capacity and to identify and eliminate bottlenecks (Narusawa & Shook, 2009). This sheet is intended to provide concrete data on how the working system operates across its entire range. In Table 4.10 an example of a process capacity sheet is presented. It should be emphasized that all the data used to feed the process capacity sheet should be provided from a process study and analysis, namely from time measurement studies. The capacity per unit time (e.g. per shift) is calculated using Eq. (4.2). In Table 4.9, the hourly capacity is considered. Moreover, the bottleneck process in this example is process number one—robotic welding (OP 20/30).

$$Capacity = \frac{Available\ Time\backslash \text{per}\backslash \text{Period}}{Total + \frac{Set\backslash \text{up}}{Change}} \tag{4.2}$$

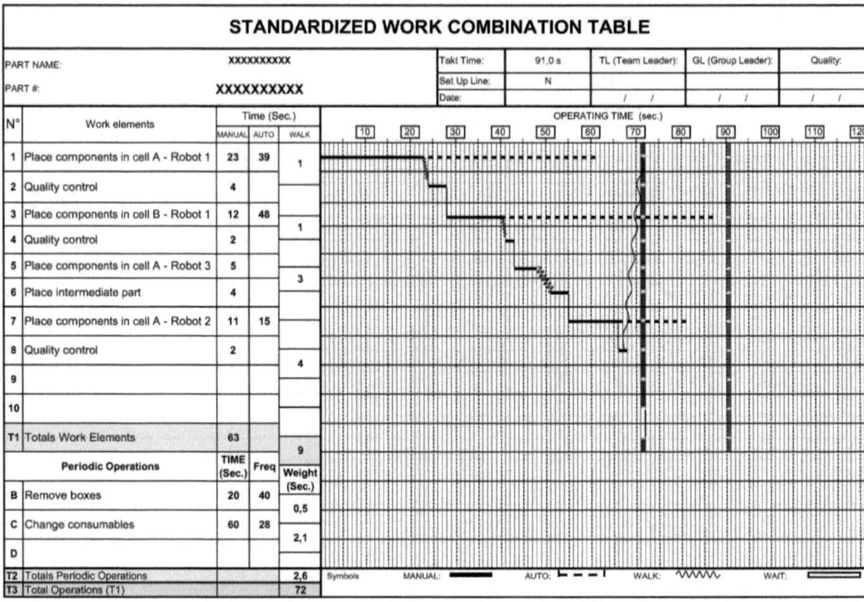

Fig. 4.9 Example of a standardized work combination table

4.2.3.2 Standardized Work Combination Table

The standardized work combination table is used to analyse jobs that entail combined work. The intention is to show the relationship in terms of time for two or more activities that occur simultaneously (Liker & Meier, 2006). It shows for each operator the combination of machine processing time, manual work time and walk time, considering each work element in a sequence, thus providing the human–machine sequence and showing the relationship between work, walk and wait time. Thus, it is very useful to identify waiting and overburdening wastes. The standardized work combination table shows visually the process lead time, helping determine the number of operators required to comply with the takt time. Figure 4.9 shows an example of a standardized work combination table.

4.2.3.3 Operator Balance Chart

The operator balance chart is a useful tool for seeing how cycle times compare to takt. It visually shows how the work elements are distributed among the different employees in a working system (characterized by multistep, multi-operator processes) in relation to takt time.

An operator balance chart is composed of vertical bars that denote the total amount of work attributed to each operator. For each operator, the vertical bar is obtained through compiling small bars that represent individual work elements, with the height of each element proportional to the amount of time required. The operator balance chart assists in the critical task of (re)distributing work elements among operators. This is a critical tool for assuring the minimization of the required number

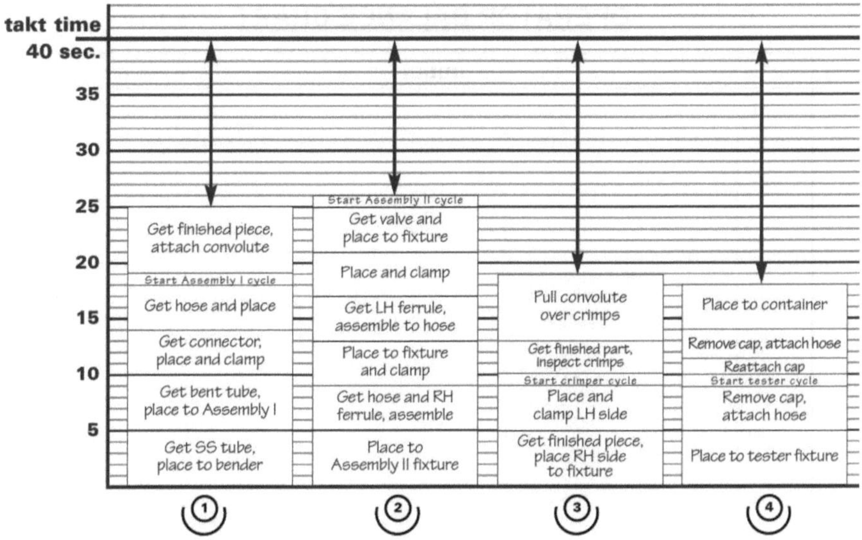

Fig. 4.10 Operator balance chart construction. Source: LEI (2008)

of operators by making the amount of work for each operator very nearly equal to, but slightly less than takt time (LEI, 2008). Data for the completion of this chart may be provided, for example, by the standardized work combination table. Figure 4.10 shows an example of the construction of an operator balance chart, while Fig. 4.11 shows the final result. The number of operators needed to meet takt time is determined by dividing the total work content by the takt time. According to Narusawa and Shook (2009), when the proportion of this calculation is less than 0.3, the number of operators should be rounded down, meaning that a certain amount of overtime will be needed unless improvements are introduced.

4.2.3.4 Standardized Work Chart

The standardized work chart is a means of identifying waste and improvements needed to achieve the desired results. It is a management tool that provides visualization of the standardized work, thus facilitating the operators' job. The standardized work chart shows operator(s) movement(s) and material location in relation to the machine and overall process layout, considering all the operations within the task under analysis. It also shows the work sequence and takt time, cycle time, work sequence and standard work in progress (WIP) (Narusawa & Shook, 2009). An example is presented in Fig. 4.12.

4.2.3.5 Job Instruction Sheet

This sheet provides the operator with a description of what he/she has to do and how to do it well, i.e. following the standardized work steps. Its main purpose is to train operators (new or experienced), allowing knowledge capitalization as all the information concerning the execution of a particular job is described in this document.

OPERATOR BALANCE CHART

Initial

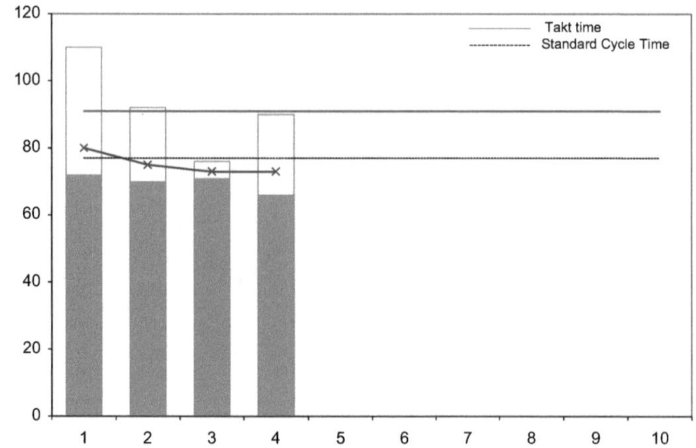

Date:

Workstation/Operator	1	2	3	4	5	6	7	8	9	10	Total
min	72	70	71	66							279
mean	80	75	73	73							301
max	110	92	76	90							368

Observations

N.° of good parts per operator per hour 51 Actual Time 70

Fig. 4.11 Example of an operator balance chart

Typically, the job instruction sheet lists the sequence of work elements within one operation. Figure 4.13 presents an example of a job instruction sheet.

There is no unified way of creating standardized work or using the documents presented above, although standardized work is commonly approached following the sequence presented below:

1. Create the process capacity sheet.
2. Build the standardized work combination table.
3. Develop the operator balance chart.
4. Create the standardized work chart.
5. Generate the job instruction sheet.

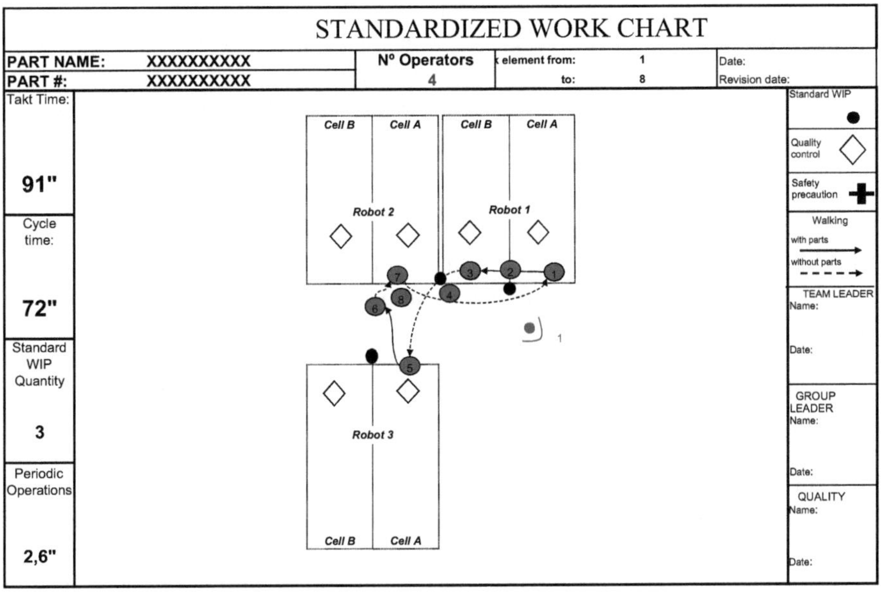

Fig. 4.12 Example of a standardized work chart

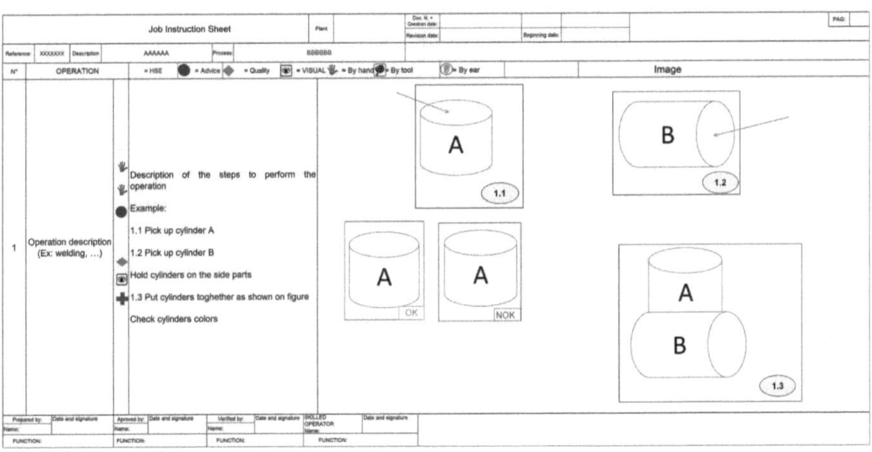

Fig. 4.13 Example of a job instruction sheet

4.2.4 Practical Application of Standardized Work

The starting point for the implementation of standardized working was the formation of a team, as in the 5S tool. The team was constituted by two team leaders and two line operators, among other individuals. Prior to the beginning of the process the team leaders were educated and trained in the standardized work principles, objectives and support documents.

As described in Sect. 4.2.2, there are three basic elements for standardized operations: takt time, work sequence and standard inventory. However, no operation is standardized until the work sequence or other types of relevant information are put on a piece of paper. Thus, during the implementation of the standardized work five primary documents were used: production capacity sheet, operator balance chart, standardized work job instruction sheet, standardized work combination table and standardized work sheet.

It was considered adequate to start the standardization of the work on one of the production lines and afterwards to spread the implementation process to other lines. In the following paragraphs, the implementation process in the selected production line is explained.

First, using the production capacity sheet, the team collected the operations times and evaluated the line capacity. The current takt time and the takt time for the following six months, considering demand projections, were also determined. It was concluded that the line had the capacity to produce 50 parts per hour. At this stage, the team leaders revealed an attitude of apprehension as it was proposed that the production level be increased from 43 parts per hour, the current production level, to 50 parts per hour within a period of six months, maintaining the same number of workers (four operators). The next step was to develop the operator balance chart to visualize the tasks distribution among the four operators (Fig. 4.14). It could be concluded that there was an imbalance in the tasks distribution. After an individual analysis of each operator's role and improvement in the work (content and distribution among the four operators), it was possible to reduce the line variability. Through the analysis of the standardized work job instruction sheet, the standardized work combination table and standardized work sheet, each operator became acquainted with his/her role, how to perform it and in how much time. Figure 4.15 presents the operator balance chart after the standardized work implementation.

The application of the standardized work led to the following results:

- The customer takt time changed from 95 s to 84 s, implying a target cycle time of approximately 72 s.
- A more efficient balance among the workstations was achieved, which allowed an increase in the hourly production level from 50 parts per hour to 60 parts per hour.
- The levels of reworking and non-conforming parts were drastically reduced.
- Through the apparent increase in new ideas for improvement suggested by the line operators, there was a clear increase in the workers' morale and accountability regarding the production process.
- The start of standardized work auditing.
- The team leader introduced several improvements in the tasks, frequently in close collaboration with the line operators.
- During the shift, the team leader now has the capability to adjust the number of line operators (n; n+1; n−1) according to the existing takt time.

Operator Balance Chart

INITIAL

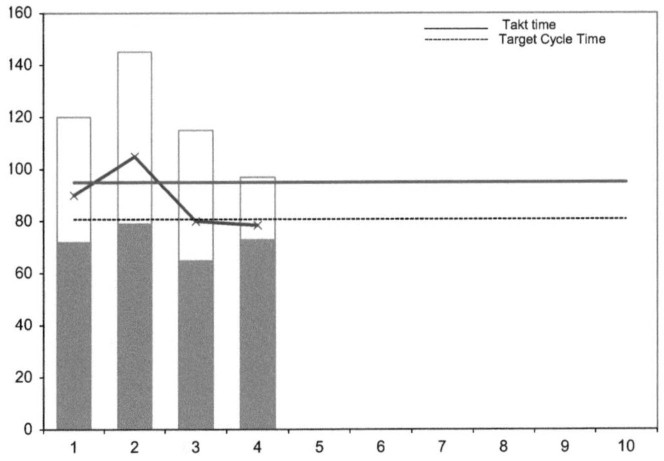

Date:

Workstation	1	2	3	4	5	6	7	8	9	10	Total
min	72	79	65	73							289
mean	90	105	80	78,5							354
max	120	145	115	97							477

Observations

- No standard operations
- High number of recoveries
- Reduced production: 43 parts/hour

Non-defective parts/person/hour 50 Real Time 72

Fig. 4.14 Operator balance initial chart

4.3 Single-Minute Exchange of Die (SMED)

The manufacturing sector is gradually being led by strong and increasing competition between enterprises competing on a worldwide stage, which in turn is forcing them to shift their focus to elements associated with innovation, competitiveness, intangible assets and the quality of their products (Vrontis, Tardivo, Bresciani, & Viassone, 2016). The competitiveness between enterprises is increasingly dependent on cost reduction and control of the engineering of production processes, with a view

Operator Balance Chart

FINAL

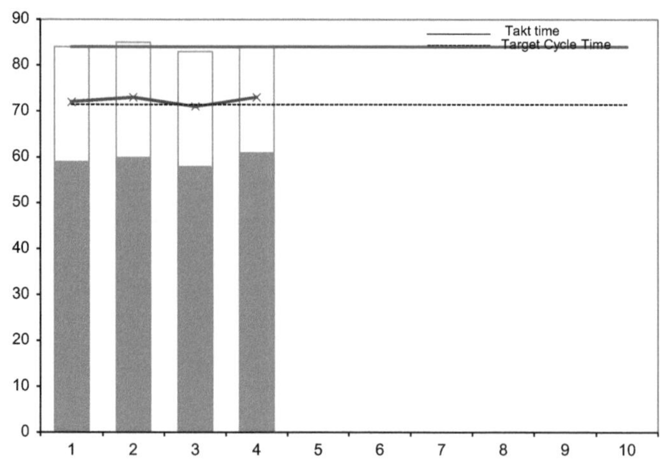

Date:

Workstation	1	2	3	4	5	6	7	8	9	10	Total
min	59	60	58	61							238
mean	72	73	71	73							289
max	84	85	83	84							336

Observations

- Production of 50 parts/hour on a regular basis
- Reduction of the number of recoveries
- Comply with work standards
- Team motivation: increase in the number of improvement ideas

Non-defective partes/person/hour 60 Real Time 60

Fig. 4.15 Operator balance final chart

to optimizing the tasks or phases of the processes that add value to the final product, which in turn the client values (Cheung, Song, & Zhang, 2017; Holtewert & Bauernhansl, 2016). Improving the understanding of processes among small and medium-sized enterprises (SMEs) can be a crucial step towards the future success of economies (de Salas, Lewis, & Huxley, 2017). Shigeo Shingo was probably the greatest contributor to modern production practices (Floyd, 2010). By applying his experience and expertise in the field of industrial engineering, he was able to assure a better quality of life for operators and for businesses (Shingo, 2017a). His theories gained traction cemented by the results obtained among the enterprises that implemented the techniques suggested (Shingo, 2017b).

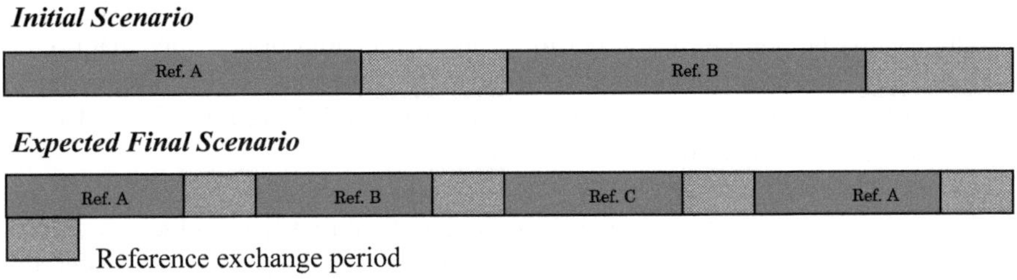

Fig. 4.16 Lot evolution with the improvement of the reference change

The SMED method can be translated into "tool change in less than 10 minutes" (Kemal Karasu, Cakmakci, Cakiroglu, Ayva, & Demirel-Ortabas, 2014). The purpose of this method is fundamentally to reduce series changing times, applying a methodology of progressive reflection that goes from the organization of the workstation to automation (Boran & Ekincioğlu, 2017). The changing of tools is defined as the time elapsed from the manufacture of the last valid part of a series to the production of the first correct part of the next series, meaning that is not only the time for the change and physical adjustments of the machinery (Carrizo-Moreira, 2014). For this method to reach its objectives, it is proposed that it is necessary to involve the different functions of the enterprise (Braglia, Frosolini, & Gallo, 2017). Thus, miraculous solutions should not be anticipated, nor should it be expected that somebody will have a brilliant idea to address the problems that it is considered difficult to resolve (Lozano, Saenz-Díez, Martínez, Jiménez, & Blanco, 2017).

In the specific case of applying the SMED method, it is often necessary to invest financial resources to achieve the desired objective, but it is usually possible to achieve considerable improvement only by setting up an operational mode with simple but effective lay-out modifications or the creation of a well-defined flow through which operators and materials can circulate (Chiarini, 2013). The schedule presented in Fig. 4.16 demonstrates the purpose of the SMED methodology (Ferradás & Salonitis, 2013).

The application of this methodology makes it possible to reduce the change times of the reference by anticipating the preparation of tools, as well as synchronizing and simplifying the tasks for every participant (Kemal Karasu et al., 2014).

4.3.1 Importance of SMED for Organizations and Its Main Advantages

In line with the JIT approach, which aims to eliminate or reduce excess stocks, strictly producing what is necessary and within the required time (García-Alcaraz & Maldonado-Macías, 2016), SMED is a methodology that can contribute most to a steady path towards achieving these objectives. As the reference change time is reduced, lots can also be reduced, which in turn leads to a decrease in stocks, making it possible to produce only what is necessary and thus enabling the organization to become more responsive and efficient (Romano, Danese, & Bortolotti, 2009).

In the elaboration and operation of the production process, substantial work should usually be put into it in order to shrink the quantity of activities that do not add any kind of value (Stefansdottir, Grunow, & Akkerman, 2017). Producing in smaller lots allows various benefits (Beemsterboer, Land, & Teunter, 2017; Kumar & Aouam, 2017; Stefansdottir et al., 2017):

- Reduction in stocks: with smaller lots, the less time they stay in the warehouse, the smaller the number of components in the warehouse and the faster the consumption of the intermediate product.
- Reduction in costs related to stocks: area, maintenance, transportation, management costs.
- Problem identification (greater reactivity): problems that arise due to high stocks are discernible; defects in production can be detected more quickly, failures can be solved through a system of limited stocks and flow and organization can be improved.

Nowadays customers are looking for precision and expect high quality, at the most competitive prices in the market and in the lowest possible time. SMED helps organizations better understand customer needs by making smaller batches but with a higher number of referrals.

SMED came into being to underline the idea that a tool change entails a considerable loss of time. With the correct application of SMED in an organization, the following benefits can be obtained (Floyd, 2010; Shingo, 2017a, 2017b):

- Flexibility—does not require a high stock of raw material due to constant reference change.
- Faster deliveries—smaller batches mean faster delivery.
- High quality—reduced time to derive the first product according to specifications.
- Increased efficiency—reduced tool change periods mean fewer inactive products.

4.3.2 How to Implement SMED

4.3.2.1 Watching

In this first stage it was necessary to use a camera to provide a recording of the change and if possible observe more than a reference change. When recording, one must take into account the number of operators and it is important to follow the operators independently. When filming starts, one should let the collaborators know why they are being recorded. There should be a focus solely on the reference change and the use of timing systems. After the observation is registered, it is necessary to decompose all the actions in terms of associated times and people.

In addition to the decomposition, it is essential that the following information panels—Table 4.11 and Fig. 4.17—are placed on the production line.

The purpose of these panels is to transmit information. The first panel provides real time information concerning what happens during the reference changes on the

Table 4.11 Evolution of reference change

FOLLOW-UP OF REFERENCE CHANGES

Date	Output reference / Entry reference	Operator's Signature	Reference Change: Starting time	Reference Change: Duration	Diagram of changing times	Comments
20-05-2007	1122/1123		08:00	15		Tuning Problems
20-05-2007	1123/1120		09:00	10		
20-05-2007	1120/1122		10:00	5		
20-05-2007	1122/1130		11.05	10		
20-05-2007	1130/1123		12:05	5		
20-05-2007	1123/1122		13:30	10		
20-05-2007	11/22/1123		12:30	15		Lift failure
20-05-2007	1123/1130		13:50	20		Lift failure
20-05-2007	1130/1120		14:45	15		Lift failure
20-05-2007	1120/1122		15:30	10		

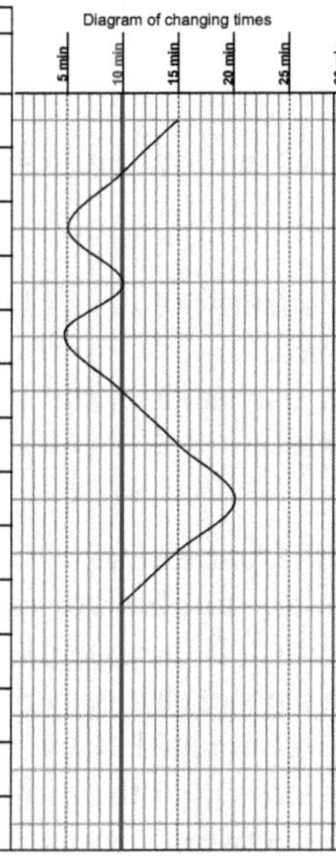

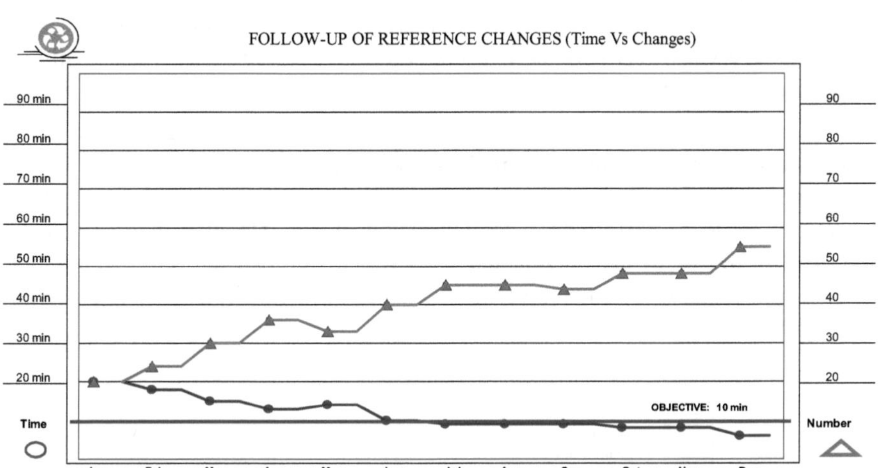

	Jan	Feb	Mar	Apr	May	Jun	Jul	Aug	Sep	Oct	Nov	Dec
TIME	20,0	18,0	15	13	14,0	10,0	9,0	9,0	9,0	8,0	8,0	6,0
Nº Change	20,0	24,0	30,0	36,0	33,0	40,0	45,0	45,0	44,0	48,0	48,0	55,0

Fig. 4.17 Evolution of the change of reference versus number of changes

line. The second provides more integrated information and an insight into the evolution of reference changes over time. When the change in reference time decreases, the number of changes must increase.

4.3.2.2 Organizing

The SMED method distinguishes two types of operations in a series change (Boran & Ekincioğlu, 2017):

- Internal operations (input exchange of die [IED]), which can only be performed with the machine immobilized.
- External operations (output exchange of die [OED]), which can be performed while the machine is running.

Identification of IED and OED Operations

For this first step, it is necessary to observe, identify and clearly record which operations are internal and external. If the external operations cannot be eliminated, they must be carried out before the tool change period. The case is similar for internal operations: if they cannot be eliminated, they must be performed at the exact moment at which the machine stops after the current series has finished.

Only this simple distinction between internal and external operations makes is possible to achieve a reduction of 30% without performing any important modification to the process by rationalizing the series change operations (Figs. 4.18 and 4.19). What usually happens is that the operator starts to perform the series change operation only when the machine is immobilized. This waiting time can easily be eliminated by improved preparation of the series change operations.

Transformation of the Internal and External Operations

After the categorization, an operative mode is created, which describes what operations are to be performed. The elaboration of this operative mode has to be described in the following documents: work instruction and the task combination table. It is important that all employees know what tasks to do and how much time they have available.

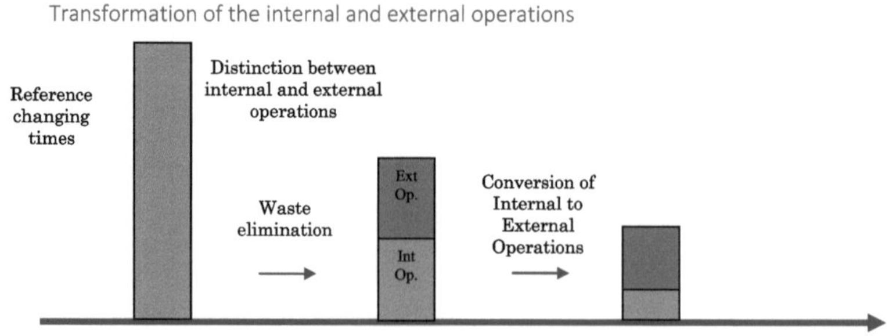

Fig. 4.18 Transition from internal to external operations

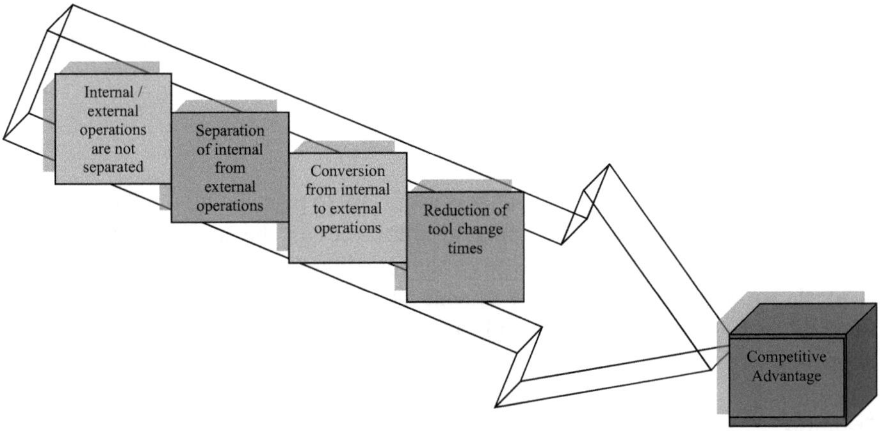

Fig. 4.19 SMED application advantages

In the elaboration of this operative mode it is important to display the following information:

- Materials needed to change references.
- Pre-check the status of the tools.
- Verification of risk functions.
- Tool transportation to the machine before the start of each change.
- Other relevant operations (preheating, etc.).

It is also relevant to accord importance to the preparation as it is a way of being able to gain more time for the reference change. If the following elements relating to external operations are safeguarded, the gains obtained are quite significant:

- Preparation of the tools and other types of assistance before starting the change of reference.
- Reduce the number of tools, standardize all fastening systems.

When carrying out internal actions, one must take into account:

- Balancing the workload of each operator involved in the reference change.
- Balancing skills to improve task synchronization.
- Having ancillary assistants to help during the reference change if the equipment imposes numerous journeys.

All actions that have to be carried out become valid only in the case that they are validated on the ground by the operators.

The checklists and operating modes should be easy to understand and whenever possible the operator should be trained to perform them. On the other hand, to facilitate the work of the operator, there should be a focus on reducing paper instructions and creating visual systems, such as those with colours or symbols.

4.3.2.3 Implementation

Communication

When an SMED action is carried out, it is necessary to post and fill out the task force launch panel and to provide training to employees in the new operating mode. Communication with employees who participate directly in the change of reference is ideal, so whenever a reference change is made in accordance with the new standard, a 5–10 min meeting should be held to analyse the aspects that can be improved.

Training

An analysis of the checklists should be made to verify if the employees have understood what was communicated to them. If it has not been assimilated, other actions should be taken to explain the new way of carrying out the change of reference.

4.3.2.4 Improving

It is necessary constantly to seek means that can facilitate the movement of moulds and tools to facilitate the work of operators and to instil the spirit of improvement among operators.

Dissemination

If the actions implemented with regard to a particular machine have a positive effect, such actions should quickly be implemented for other machines of the same family.

The main steps of the SMED application method can be seen in Fig. 4.19 and are as follows:

1. Separate internal operations from external operations.
2. Transform internal operations into external operations.
3. Reduce internal operations and external operations.

4.3.3 Practical Application of SMED

The methodology implemented is divided into two parts. The first refers to the method on which the SMED methodology is based, which is divided into five points. In the second part are the results obtained from the application of the methodology.

The method analysis comprises the following steps:

Step 1—Supply and exchange of materials
Step 2—Tool preparation
Step 3—Modification of the production tool

Step 4—Control and acceptance of the product
Step 5—Synchronization of operations

Step 1—Supply and exchange of materials

The supply and exchange of raw materials generates many handlings and the raw material goes a long way before reaching its end; some incidents may occur.

Step 2—Tool preparation

The preparation of tools should be performed by specialized persons who must follow standard procedures that are common to several people or processes.

Step 2.1. Standardization of tools. The use of standardized methods eliminates many unnecessary movements that lead to wasted time. Standardization is paramount; in addition to reducing many unnecessary movements, it also leads to a reduction in internal and external operations, increasing operator autonomy.

Step 2.2. Specific means of movement. Usually the means that are used for the reference change are also used for performing other operations and are not always available when needed. Thus, where possible, specific means should be created, not only for the change of reference but also for all operations.

Step 3—Modification of the production tool

Modification of the tool is performed by changing the tool or shaping the machine elements. The final position is usually obtained after making adjustments. To reduce the final modification time of the tool, it is necessary to go from the adjustments to production without using "zero keys".

Step 3.1. Analysis of the adjustment. The adjustment consists of obtaining, through successive approximations, the definitive position of the tool or of some element of the machine.

Step 3.2. Integrated placement. Integrated placement is simply putting the elements in pre-defined positions. If a part, for example, needs to be tightened, instead of being executed with a key appropriate to the operation, a threaded fastening device is placed in the thread itself, thus eliminating the need for a key.

Step 4—Product control and acceptance

Control points, acceptance limits and control means are determined at the time of product creation and are never altered or improved.

As the use of the SMED methodology is considered, these points are also analysed because:

- the acceptance time is also part of the reference exchange;
- reducing the lot size increases the performance of these operations.

Step 5—Synchronization of operations

The synchronization of operations is very important since if the operation of exchange of reference is carried out by only one person, useless displacements will be implemented; when the same operation is carried out by two or more people, these same movements are reduced. However, each operator must know exactly what his/her function in the reference change is, thus avoiding the occurrence of problems.

Fig. 4.20 Time applied to reference change

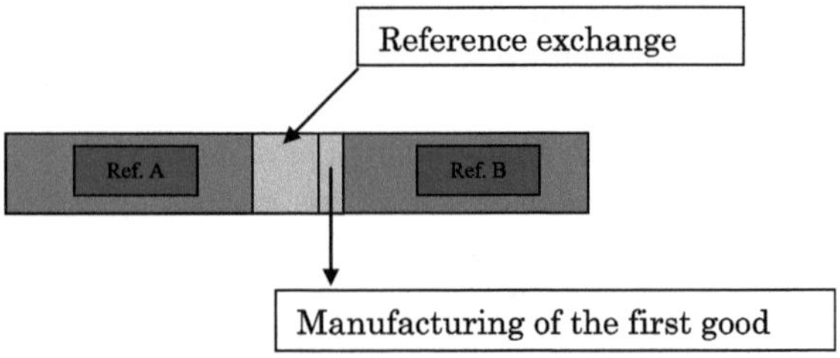

The reference exchange time is the elapsed time between the last good piece of the previous reference to the first good piece of the new reference, as illustrated in Fig. 4.20.

4.3.3.1 Analysis of the Exchange

The analysis of the exchange operations involves the following points:

- Continuous analysis of the tool change process.
- Conducting a work study.
- Interviewing workers.
- Making videos for later analysis and review.

Start of the Project

First, a time-out was carried out to check the buffers and to gain a clearer perception of how the reference change operation is performed. It turns out that this is a rather time-consuming operation and takes great technical precision, requiring highly qualified and experienced people to perform it. Moreover, one of the greatest difficulties concerns the weight of moulds, in the order of 10 tons, with an injection pressure of around 900 tons.

From observation, there is some concern related to the reduction of time in the change operation; however, there is no concrete plan for optimization and if any new idea arises, there is no support to describe it or for it to be of later use to other exchange teams. One of the points that was considered initially, but did not materialize, was related to the instant of mould change and the fact that the new mould would not be found in a location near the change operation. What actually happens is that the mould to be replaced remains in operation until the new mould is ready for replacement. A significant reduction in the reference change time was observed in this operation.

However, some ideas emerged that could become relevant:

- Initially, the skills and skill gaps of reference change teams should be analysed in order to reshape and balance those teams.

In each team there must be a "Team Leader", whose function is to lead the team, being responsible for the detection of problems within the team and the training of the team members, as well as communicating any optimizations and improvements

Table 4.12 Operations used initially for the reference change

SMED Application Sheet		Time	Operations	
No	Activity description	Seconds	Internal	External
1	Turn off all machine circuits	30	X	
2	Pick up and open toolbox	12		X
3	Remove gas, water and electric hoses	600	X	
4	Fasten mould to bridge	30		X
5	Unscrew the mould	600	X	
6	Evacuate the mould	180		X
7	Clean the injection nozzle	120		X
8	Change bridge from one mould to another	120		X
9	Provision the new mould	180		X
10	Place centring washers in the mould	15	X	
11	Loosen and raise/lower fixing point	60	X	
12	Withdraw from the greenhouse raw material from previous production	360		X
13	Place raw material in the greenhouse for the new production	360		X
14	Withdraw the hand held from previous production	240	X	
15	Place the hand caught in the new production	240	X	
16	Withdraw material from previous production and place material for the new production	180		X
17	Put new mould in holder and centre	180		X
18	Measure the distance from the mould to the machine axis	120		X
19	Adjust the position of the mould, fit and fasten mounting plates	360	X	
20	Place new hoses and electrical plugs	600	X	
21	Identify problems with hoses, look for, bring and fit new hose	120		X
22	Connect all machine circuits	30	X	
23	Floppy disk placement for machine tuning	120	X	
Total		4857	11	12

made by the team to other teams, so that everyone has access to the maximum amount of information.

- A checklist (Table 4.12) must be created for all material needed to perform this operation. There should also be an operating procedure so that when the change of reference is made, the person in charge of the preparation of the material knows exactly what material is required.
- One of the problems analysed during the mould changing operation concerned the connecting hose. This problem is ostensibly frequent, resulting in a loss of time in the operation verifying where another is located to replace the non-compliant item; an operator moves to the material store, inserts the reference in the computer, finds the new hose and finally places it in the mould.

With the existence of the checklist, this problem could also be easily solved, since whenever a change of reference happens, a mention of the replacement of hoses should be included in the "checklist".

- Another problem analysed was the constant measurement process as the mould should not have a very high deviation, so measurement tools included in the mould itself should be made available or guides should be developed to facilitate the correct placement of the mould.
- The capacity of greenhouses should also be re-evaluated as certain materials require preheating and/or moisture removal. This operation is time-consuming and can take up to 4 h; in most cases, the reference change is complete but the mould is still in the greenhouse for a long period of time.

4.4 Results

Of the suggestions presented, two are in the implementation phase, the rest being in the study phase for future implementation.

The first to be implemented was the checklist. One existed, as seen in Table 4.10, but it was not up to date and its use was poor. The one being proposed and developed goes beyond the existing one, considering the material necessary for the reference change, the material with which some kind of problem constantly occurs, thus creating a kind of cart where all the material necessary to carry out the operation can be found. The proposed checklist can be seen in Table 4.13. Another problem resolved concerns the connecting hoses. Currently, when a reference change of a certain mould is initiated, an operator is responsible for the presence of a hose kit, allowing replacement in the event of a problem. A study is also being carried out on the capacity of greenhouses, aiming to define more efficient planning so that the temporal distance between the change of reference and the raw material being ready to be placed in the mould is as small as possible.

From the analysis carried out in the field, the need for greater organization is indicated due to the inclusion of the material cart. Thus, whenever a tool is needed, a simple displacement is sufficient compared to the distances previously travelled. The greatest gain is related to the connecting hoses, as the time interval spent by the operator moving to the warehouse has been eliminated and there is now a replacement kit available at all times.

4.5 Kanban System

Many organizations have developed techniques, methods and systems to make their operations more productive and efficient with a view to achieving better performance and becoming more competitive. One system that has been chosen by organizations to attain these goals is Kanban. According to Surendra, Yousef, and Ronal (1999), the widespread implementation of the Kanban system is based on the

Table 4.13 Operations currently used for reference change

SMED Application Sheet		Time	Operations	
N°	Activity description	Seconds	Internal	External
1	Turn off all machine circuits	29	X	
2	Pick up and open the toolbox	15		X
3	Remove gas, water and electric hoses	550	X	
4	Fasten the mould to the bridge	30		X
5	Unscrew the mould	590	X	
6	Evacuate the mould	190		X
7	Clean the injection nozzle	100		X
8	Change bridge from one mould to another	135		X
9	Provision the new mould	185		X
10	Place centring washers in the mould	20	X	
11	Loosen and raise/lower fixing point	55	X	
12	Withdraw from the greenhouse the raw material from previous production	355		X
13	Place raw material for new production in the greenhouse	350		X
14	Withdraw the hand held from previous production	250	X	
15	Place the hand caught in the new production	240	X	
16	Withdraw material from previous production and place material for the new production	170		X
17	Put new mould in holder and centre	180		X
18	Measure the distance from the mould to the machine axis	135		X
19	Adjust the position of the mould, fit and fasten mounting plates	360	X	
20	Place new hoses and electrical plugs	600	X	
21	Connect all machine circuits	25	X	
22	Floppy disk placement for machine tuning	120	X	
23	Turn off all machine circuits	29	X	
Total		4684	12	11

benefits that companies recognize through its implementation, such as cost savings, developing flexible workstations, reducing waste and scrap, minimizing waiting times and logistics costs and reducing inventory.

The Kanban system represents a pull-type production/inventory control policy that aims to control the production facility with respect to actual demand, rather than forecasts. It has gained considerable attention among the industrial engineering community. Kanban (kahn-bahn) is a Japanese word that means "visible record" or "visible part" (Surendra et al., 1999).

Kanban is considered an information system that helps control the production mix, necessary quantities and lead times in every process (Monden, 1983) in the context of JIT production. This system is implemented in manufacturing systems in which there is in-process inventory between work-stations, which represents a risk of station blocking (Berkley, 1992). Kanban is a lean production technique that aims to eliminate labour and inventory waste through the pull production model, according to which production starts only after receiving a customer order. That is, instead of forecasting the production of a specific item, Kanban produces items in direct relation to market demand. The main rule of Kanban is that production only starts or items are only moved after receiving the signal from the customer to do so (Surendra et al., 1999). The idea underpinning Kanban is that workstations produce or deliver desired items only when needed, thanks to a visual signal in the form of the reception of a card, box, or empty container (Sugimori, Kusunoki, Cho, & Uchikawa, 1977). As a lean tool, Kanban it is implemented on the shop floor to align material flow by removing all sources of waste. To identify inefficiencies and assist production, when implementing Kanban in the manufacturing area, it should be supported by a set of tools, such as Heijunka board, lot formation post, Kanban chute and Kanban post (Kumar & Panneerselvam, 2007; Singh, Garg, Sharma, & Grewal, 2010).

The Kanban system contributes to cost reduction not only by eliminating waste, but also by being more responsive to change, facilitating quality control and giving importance, trust and support to the employees running the processes (Chalice, 2007).

The Kanban system works effectively under specific production and market conditions (Lage & Godinho Filho, 2010), such as: (1) regular demand from the customer to make it easier to hold Kanban stocks in the traditional supermarket style; (2) low product variation to guarantee a low quantity of inventory; (3) clear flow—facilities that are organized in a silo style with all similar processing being done in one location are hard to control using a Kanban system. A better arrangement is one in which all processes are organized together to provide a flow line or cell; (4) small dedicated machines used within product flow lines; (5) quick changeovers—avoiding large batches and bottlenecks within production by using SMED techniques; (6) repeatable and reliable processes using TPM, 5S, operator-driven quality improvements and standardized operations; (7) reliable suppliers to support the Kanban processes.

Since these conditions are not observed in all organizations, the Kanban system has some limitations reported in the literature (Aggarwal, 1985; Grünwald, Striekwold, & Weeda, 1989; Monden, 1984; Sipper & Bulfin Jr., 1997) such as: not being adequate in situations with unstable demand, processing time instability, non-standardized operations, long setup times, a great variety of items and uncertainty with regard to the supply of raw material. Also, to ensure the success implementation of Kanban, good inventory management is mandatory, vendors and suppliers should be involved, the quality should be improved and employee and top management should be engaged (Kumar, 2010). In this system, each station operator should keep pace with Kanban's continual "pull", since inventory is continuously moving. Also, the suppliers should be pre-certified by quality assurance, which

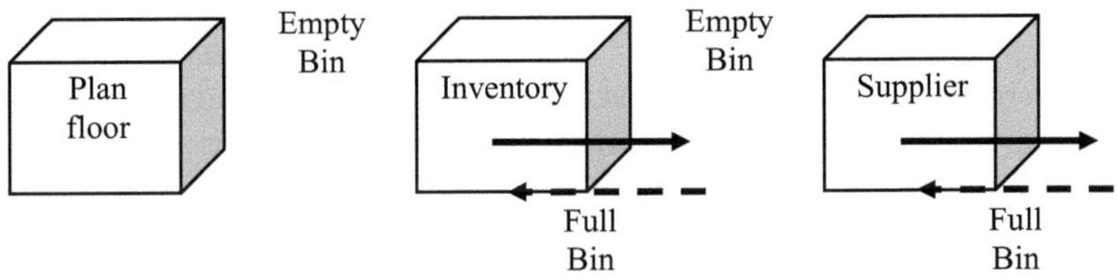

Fig. 4.21 Kanban three-bin system

allows them to move supplies to the production process in a "ship-to-use" programme rather than ship to the warehouse.

4.5.1 Implementing a Kanban System

To illustrate the implementation of a Kanban system, the example of a three-bin system (Row, 2017) is used. In this kind of system three bins are used to store the items in the production area, to hold items in the inventory department and also to deliver them to suppliers:

Bin 1—is available on the floor of the production unit using the product.

Bin 2—is available at the inventory department of the factory, where the production staff obtain raw materials.

Bin 3—is available at the premises of the supplier selected to deliver the materials.

Each bin contains a Kanban card with detailed information on inventory and the date on which stock was received. During the production process, the materials from the bin are used and once it depleted, the Kanban card is returned to the inventory department (Fig. 4.21). Next, the inventory department replaces the empty bin with a full bin obtained from the supplier or vendor and sends the empty one to the vendor or supplier for replenishment of materials. This three-bin method does not require on-site storage of materials until they are required.

The main benefits associated with the implementation of the Kanban system include reduced inventory holding, improved material flow, elimination of overproduction, better control of material handling, development of visual scheduling and process management, enhanced response to the market, minimization of obsolete inventory, better supply chain management (Groos & Mclnnis, 2003) and the minimization of waste (Giard & Mendy, 2008).

The Kanban system is a mechanism for managing and controlling the flow of material in manufacturing. The implementation of a successful Kanban system requires some conditions, such as the following:

- *Regular demand from the customer*—Irregular demand from customers makes it difficult to predict and consequently it can be hard to hold Kanban stocks in the traditional supermarket style.

- *Low product variation*—If many different products are made, the amount of holding stocks could be too large. This burden should be reduced by ensuring that there are many commonly used parts between products and a postponement strategy is used to make the product unique as late in the process as possible.
- *Clear flow*—The layout of facilities is crucial to the success of Kanban. Facilities that are organized in a silo style are hard to control with a Kanban system. A better layout is one in which all processes are organized together to provide a flow line or cell.
- *Small dedicated machines*—Large machines driving the use of large batches represent a bottleneck for the facility. It is better to use smaller dedicated machines within product flow lines.
- *Quick changeovers*—The setup times of many machines and processes are too long. This can result in large batches and create bottlenecks within production. To avoid this potential problem the SMED technique can be used.
- *Repeatable and reliable processes*—If machines are constantly breaking down and processes are not repeatable, the Kanban system alone cannot help. In this context, the use of TPM, 5S, and standardized operations jointly with Kanban would be the better option.
- *Reliable suppliers*—Suppliers are a vital part of the process and it is important to ensure that they can support the Kanban process in a reliable way.

A Kanban card is used to regulate the flow of materials throughout the processes. The Kanban card is driven by downstream need and triggers upstream production. Upstream production is then initiated to replenish those parts that have been withdrawn (Singh & Shek, 1990).

In a production environment that is controlled by Kanban, the material flow and the information flow run in contrary directions. The material flow is from the preceding to succeeding stage, but the information flow is downward and from the succeeding to preceding stage. The production starts only when workstations receive a card. Empty containers give the signal that more parts are needed for production. In this process, each workstation only produces enough items to fill containers and then stops (Shahabudeen, Gopinath, & Krishnaiah, 2002) (Fig. 4.22).

According to Klipp (2012)—and considering a context in which a process or service must be managed—there are six important steps that should be performed prior to the implementation of a Kanban flow: (1) map the workflow, (2) visualize the WIP, (3) set initial WIP limits, (4) get Kanban working, (5) look for bottlenecks and (6) inspect and adapt.

1. Map the workflow

   In this phase the major processes in the department or organization must be identified, as much the steps in the individual processes. It is important to understand the origins of the tasks, how they are prioritized, defined and assigned and what steps it takes for the work to be completed and done right. This represents the workflow.

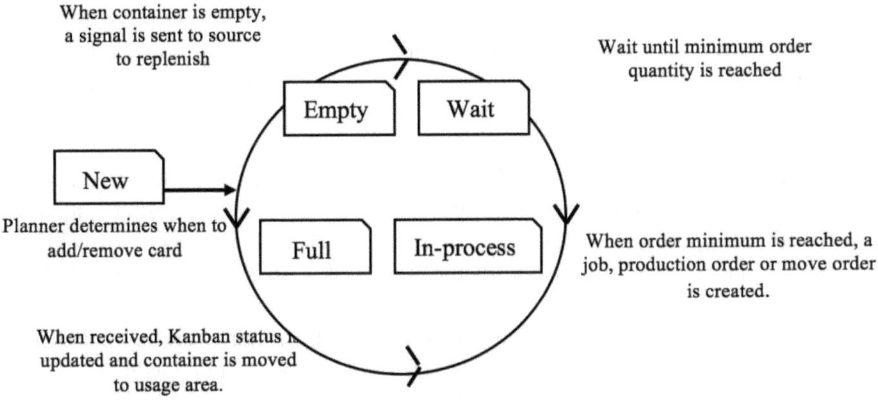

Fig. 4.22 Kanban card flow

It is important to understand that documenting the steps of a process should be done in line with the existing process and not based on the ideal process. The starting point should be the process that exists in the real world. If there are inefficiencies, bottlenecks, or missing or unnecessary steps, they will come to light and the workflow can be modified. Then, for each step, the variety of tasks that could be done at once must be considered.

Once the workflow is mapped, a Kanban board is built. The Kanban board comprises a table that has one column for each step in the workflow. The table can be drawn on paper, on a whiteboard, on a corkboard, or using an online tool.

2. Visualize WIP

   Once the process is mapped and the Kanban board has been created, tasks are identified on it. Tasks represent something that must be done and should be linked to a suggested name. Also, depending on the task and the process of which it is part, the Kanban board should show or track other information, such as the creation date, deadline, created by, priority, task type, description, notes, definition or requirements for "complete/finished" and history.

3. Set initial WIP limits

   Initially, it is difficult to identify the ideal workload for various tasks or processes because setting low initial limits can be painful, but will yield faster results, while setting high limits initially and purposefully lowering them over time may have slower results, but can help ensure buy-in and adoption within the team.

   A recommended practice in setting initial WIP limits is to involve the team in documenting and defining the process, as well as setting WIP limits for each phase. The team must understand that limiting the number of tasks they are working on contributes to reducing the time it takes to get any one task done and improves quality by placing greater focus on fewer tasks. This approach leads to improved efficiency as things not only get done faster, but are also better than ever before.

Once the limits of WIP are defined for each phase of the process, they are written above the columns on the Kanban board that represent steps in the process. This makes it possible for everyone to be aware of the limits and they should try not to exceed them.

4. Get Kanban working

   Kanban is a "pull" system, which means that one stage of the process pulls work from the previous stage. This approach contributes to reducing the quantity of WIP, as opposed to a "push" system in which each stage works as quickly as possible and then pushes work to the next task—no matter how much WIP already exists.

   Attending to this system, when someone is ready to start working, he/she has only to look at the board to identify the tasks to be done and pull the next task into the column representing the next step in the process. The tasks assigned to specific people become their responsibility until finished and someone else pulls them into the next step.

5. Look for bottlenecks

   Once a process is mapped and the Kanban board is set up, an illustration of each task is performed and the position of every task in the workflow can be seen. This makes it possible to see bottlenecks forming early. If WIP limits are set correctly, there will be a limit just as a bottleneck begins to form.

   Part of the power of Kanban is that it makes it easy for everyone to see a problem forming and a solution can be found before a bottleneck is identified. This avoids incomplete work and people starting to jump from task to task without information on the next tasks that should be performed, increasing the complexity of the overall workload.

6. Inspect and adapt

   Many advantages associated with the Kanban system come simply from visualizing workflow, but others need more powerful tools. Two powerful tools for tracking improvements concern the assessment of the lead time and cycle time. Lead time essentially represents how long it takes to get something done, from the time someone asks for something until they receive it. Cycle time is how long it takes someone to finish a task once they have started it.

   A software tool can be used to compare metrics at various points in the process and see the results of process improvements. Collecting and using these data make it possible to determine the delivery date and the time necessary to perform a certain task. These metrics should periodically be revised by the team to improve their performance.

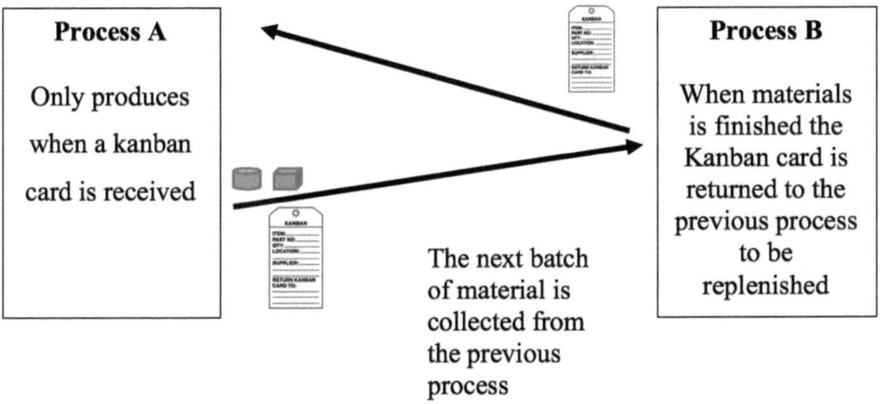

Fig. 4.23 Single-card Kanban system

4.5.2 Types of Kanban Systems

The Kanban system is established on the shop floor to align the flow of materials and allows the removal of all sources of waste. According to Kumar and Panneerselvam (2007), waste is anything that the customer considers to have no value and is not willing to pay for. Wastes can be categorized into the following types: transportation, over-inventory, excess of motion, waiting, over-processing, over-production and defects.

There are various types of Kanban that can be implemented depending on the specific characteristics of companies and production flows. In the literature, some of the Kanban systems that can be found are identified by Lage and Godinho Filho (2010): the Modified Kanban System (MKS), the Independent Kanban Control System (IKCS), the Generic Kanban System (GKS), the Flexible Kanban System (FKS), Job-Shop Kanban, e-Kanban, Kanban cards, the Kanban bin system and CONWIP (CONstant Work In Progress) systems.

Another type of Kanban system focuses on the number of cards associated with each item. In this regard, Agus and Hajinoor (2012) consider that there are two types of Kanban system: a single-card Kanban system and a two-card Kanban system. The single-card Kanban system uses only one type of Kanban card to trigger upstream production when needed. This card is called production Kanban (P-Kanban) and specifies the kind and quantity of product which the preceding process must produce.

In a single-card Kanban system, parts are produced and bought based on a daily schedule and deliveries to the user are controlled by a withdrawal Kanban (W-Kanban), which specifies the kind and quantity of product a manufacturing process should extract from a preceding process (Fig. 4.23). The single-card system is a push system for production coupled with a pull system for delivery to the point of use. Single-card Kanban controls deliveries in a very efficient way, so that the work centre never has more than a container or two of parts and the stock point serving the work centre is eliminated. The single-card system is suitable for

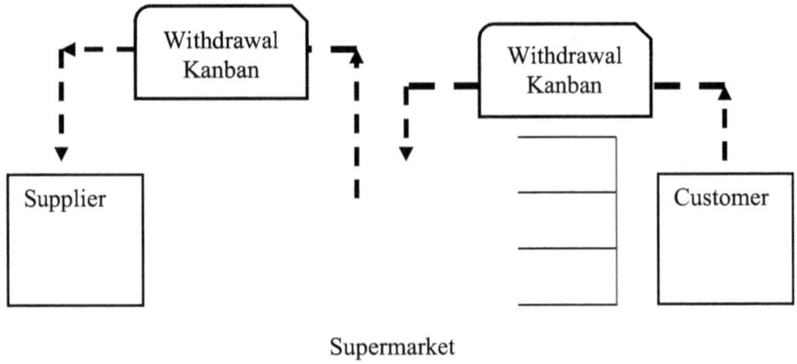

Fig. 4.24 Two-card Kanban system

companies with a relatively small range of end products, or products that are not subject to rapid, unexpected changes in demand levels, it being easy in such cases to associate the required quantity and timing of component parts with the schedule of end products (Monden, 1993).

The two-card Kanban system uses a W-Kanban card and a P-Kanban card (Fig. 4.24). When a process finishes using the materials to which the Kanban card is attached the card is returned to the previous process. This then gives the previous process an order to manufacture replacement parts. In a multiple card system, the process will typically have to wait for a set number of cards to be returned before starting to manufacture the next batch. A P-Kanban is attached to the buffer stock after each production process; it is removed when the part is extracted by the downstream stage and is used to initiate the production of new items (Mazza, 2015).

Other types of Kanban systems, such as supplier Kanban, signal Kanban, express Kanban, emergency Kanban, through Kanban and common Kanban, can be found within the TPS (Monden, 1998a, 1998b, 1998c). Huang and Kusiak (1996) have further grouped the different types of Kanban together into primary Kanban, supply Kanban, procurement Kanban, subcontract Kanban and auxiliary Kanban.

Unfortunately, the traditional Kanban system is not adequate in situations of unstable demand, processing time instability, non-standardized operations, long setup times, a great variety of items and uncertainty with regard to raw material supply (Lage & Godinho Filho, 2010).

The most common problem with paper-based Kanban system is lost cards, which leads to material outages, waiting time, extra costs and eventually a lower service level (Drickhammer, 2005). Conventional Kanban systems lack the ability to track and monitor physical cards (Wan & Chen, 2008). Given the rapid advance of the Internet and web-based technologies, web-based Kanban has been developed. This kind of Kanban has various advantages, such as minimizing human error, real-time tracking, performance monitoring, greatly enhanced information content and instant delivery of Kanban cards (Wan & Chen, 2008). The major benefit of the web-based system is flexibility; therefore, it is more suitable for current competitive environment.

To overcome the problems associated with traditional Kanban, data can be obtained through Internet-based computing, using e-Kanban to provide greater coordination of the flow of goods. This also presents an answer to the new environmental challenges that are affecting companies. The use of the Internet has enabled companies to compete in domestic and international markets and has improved responsiveness. Employing information and communications technology (ICT), such as radio frequency identification (RFID), electronic data interchange (EDI), enterprise resource planning (ERP) and manufacturing resource planning (MRP) systems, may improve the efficient management of supply chains. With such technological developments, the Kanban concept has also been enhanced with the development of e-Kanban. It is mostly an Internet-based application that automates the transfer of the Kanban signal via a barcode scan to an upstream supplier (Barkmeyer, 2007). According to Lage and Godinho Filho (2010, p. 13) "e-Kanban is a variation of Kanban with modifications". Therefore, the essential characteristics of Kanban can also be applied to e-Kanban, although the implementation could be more difficult.

In e-Kanban, when the bar code on the Kanban card is scanned, the status of the container changes from "empty" to "in process" to "in transit" or to "full" (Drickhammer, 2005), depending on the situation. When a lot size is extracted from the Kanban buffer, the bar code on the container is scanned and the Kanban status changes from "full" to "empty". A new electronic Kanban, representing a new order, is generated in the system and sent to the supplier by a secure website or email. From the time of receipt of the signal, the status changes to "in process" and the supplier needs to prepare the new material order within a set time limit.

The advantages of e-Kanban over manual Kanban are manifold, as can be seen from Table 4.14.

4.5.3 Main Kanban Rules

In the literature various rules have been used as guidelines for the implementation of Kanban systems. Schonberger (1983) argues that the simplicity and effectiveness of the Kanban system should be based on the following rules:

1. No parts may be made unless there is a P-Kanban authorizing production. Parts production comes to a halt rather than making parts not yet asked for. This represents a pure pull system, in which workers may do maintenance or work on improvement projects when there are no P-Kanban cards in the dispatch box.
2. There is precisely one conveyance Kanban (C-Kanban) and one P-Kanban for each container and the number of containers per part number in the system is a carefully considered management decision.
3. Only standard containers may be used and they are always filled with the prescribed quantity—no more, no less. With such careful control of quantities per container, as well as the number of containers per part number, inventory control is simple and far more precise than manual or computer-based systems.

Table 4.14 Advantages of e-Kanban over manual Kanban

Advantages of e-Kanban over Kanban cards	Author(s)
Manual card handling and order-entry activities are reduced to a minimum, with no chance of lost cards. This simplifies the purchase process and releases the administrative workforce.	MacKerron, Kumar, Kumar, and Esain (2014)
The replenishment signal cannot get lost, which almost eliminates the risk of starving.	Ardalan and Diaz (2011), Drickhammer (2005), Krishnaiyer and Padmanabhan (2009)
Formalizes the communication process and eliminates many of the manual errors which arise from faxing Kanban orders or emailing spreadsheets to suppliers. This also overcomes physical distance, a limitation of the use of original Kanban.	Hadaya and Cassivi (2007)
Improved flow coordination of the purchased materials, reduction in WIP, automatic record keeping, increased flexibility in supply and an increased and faster responsiveness to fluctuations in demand.	Lage and Godinho Filho (2010)
Communication with suppliers is clarified: e-Kanban acts as a "control panel", enabling real-time visibility of demand signals and providing an overview of the status of every Kanban in the system.	Lage and Godinho Filho (2010)
Opportunity to ease and speed the analysis of supplier performance. Reporting on system performance is also both faster and more accurate.	MacKerron et al. (2014)
e-Kanban is more efficient in supporting activities in production, providing wider application and easy integration with other production planning and control software, such as ERP systems, allowing integration across networks.	Bendavid, Boeck, and Philippe (2010), Ultriva Inc. (2007)

In Kniberg (2009), three rules are highlighted for Kanban system implementation:

1. Visualize the workflow.
2. Limit WIP at each workflow stage.
3. Measure the lead time (i.e. average time to complete one item).

The production sequence at a station is decided according to the priority rule chosen to support this decision. The sequence is updated as new information concerning the station's raw material availability and P-Kanban cards become available. When the station completes a job, the next product to be produced is decided according to the priority rule. Numerous studies have analysed the performance of priority rules in workshops and have concluded that the shortest processing time (SPT) is the one that presents superior performance in a variety of environments (Blackstone, Phillips, & Hogg, 1982).

In a conventional JIT system, workstations use the priority rule first-come, first-served (FCFS) to select the item to be produced next from the available P-Kanban cards (Berkley & Kiran, 1991). As the system's environment and operational characteristics become more complex and dynamic, additional information may be used to determine the production sequence and thus improve system performance (Ardalan, 1997).

As the Kanban system influences the level of inventory, it is important for managers to determine in the initial phase of the system implementation the number of Kanban cards used (Fukukawa & Hong, 1993). The total number of Kanban cards must be determined with accuracy based on the inventory required to achieve smooth circulation of Kanban cards in the system.

Under the continuous improvement philosophy, the number of Kanban cards has been identified as an important variable in JIT systems, being directly related to the control of inventory levels and customer delivery lead times. The number of Kanban cards also affects a variety of costs, including inventory, quality and material handling (Miyazaki, Ohta, & Nishiyama, 1988) and customer service. A reduction in the number of Kanban cards when there are large variations in demand and processing times (Huang, Rees, & Taylor, 1983; Lee & Seah, 1988) results in trade-offs between inventory costs and customer wait times (Ardalan, 1997).

Before determining the number of Kanban cards, it is important to analyse and determine the recommended length of the withdrawal cycle, i.e. the time between two consecutive movements of withdrawal Kanban cards by a station to upstream stations. Short withdrawal cycles improve the frequency of communication among stations (Hall, 1983; Shingo, 1981), resulting in a higher inventory position in the input stock points. Therefore, shorter withdrawal cycles result in a lower possibility of station stock-outs. Short withdrawal cycles also lead to the withdrawal of completed items from the output stock points of stations and thus a smaller possibility of blocking. A station remains idle when there are no P-Kanban cards in the station. Thus, shorter withdrawal cycles result in higher inventory levels in the system (Karmarkar, 1986), making it more responsive to customer demand. The withdrawal cycle should be carefully determined because of its effect on system performance (Kim, 1985; Miyazaki et al., 1988).

To operate a generic Kanban system, the number of Kanban cards and lot sizes used in the systems must be determined. The number of Kanban cards and lot sizes directly affect system performance (WIP and cycle time) (Chan, Yin, & Chan, 2010).). There are trade-offs between WIP and cycle time and decision making concerning optimal performance is not easy.

There are several ways of determining the correct number of Kanban cards that should be used in a JIT system, such as simulation studies, mathematical programming models and queueing and Markov chain models (Akturk & Erhun, 1999; Berkley, 1992; Huang & Kusiak, 1996; Sendil Kumar & Panneerselvam, 2007). Simulation can be used to test different configurations and thus find the optimal number of Kanban cards that contribute to superior performance (Davis & Stubitz, 1987). Another approach used to determine the optimal number of Kanban cards is mathematical programming models. Kimura and Terada (1981) developed the first mathematical model, assuming that there are no capacity limits on production, to determine the number of circulating Kanban cards by formulating basic balance equations for multi-stage serial production setting with a single part type. Finally, some papers propose approaches based on queueing and Markov chain models to optimize the number of Kanban cards. Wang and Wang (1991) present a continuous time Markov model to determine the number of Kanban cards in a multi-item, multi-stage, dual-card Kanban system with a single withdrawal Kanban at each stage to minimize the costs.

A simpler method for determining the number of Kanban cards can be found in Monden (1983). This author observed how the Japanese set the number of Kanban cards at a work centre and described it using the following equation:

$$Number\ of\ kanbans \geq \frac{demand \times lead\ time \times (1 + safety\ factor)}{container\ capacity} \tag{4.3}$$

where

Lead time = processing time + waiting time + conveyance time + Kanban collecting time (Monden, 1983, p. 70).

Kanban collecting time = time during which Kanban cards are waiting to be picked up or returned to the beginning of a production operation.

If the safety factor in Eq. (4.3) is set equal to zero and if demand is expressed in containers per unit time, the formula becomes:

$$Number\ of\ kanbans \geq demand \times lead\ time \tag{4.4}$$

This being so, the number of Kanban cards is at least the lead time demand expressed in terms of containers. This rationale, supporting the determination of the number of Kanban cards, only allows enough buffering to compensate for lead time demand.

4.5.4 Kanban and Production Levelling

The TPS is an example of a pull system that aims to improve the efficiency of manufacturing operations by eliminating waste. The TPS includes, inter alia, the following elements: the Kanban system, production smoothing (called *heijunka*), shortening setup time and standardization of operations (Monden, 1998a, 1998b, 1998c; Ohno, 1988a, 1988b, 1988c). In the TPS, a pull system tends to be used to coordinate the stages of production, which means that only the quantity required by the customer is produced.

Among the elements of the TPS, *heijunka* is used in combination with the Kanban system (Liker, 2004; Monden, 1998a, 1998b, 1998c) and describes the effort to match the workload to the capacity or capability of machines and operators. Build-to-order and build-to-stock companies often lack a balance in production quantity. This causes hidden problems and poor quality because of the unbalanced use of resources. In addition, there is an uneven demand on upstream processes, which causes a bullwhip effect (Coleman & Vaghefi, 1994). In this context, the levelling of production is essential. The levelling of production requires that to produce only what the customers need, the supplier has to adapt the production quantity to the customers' orders and produce to takt time. The takt time represents the synchronization between production and sales. It is calculated by dividing the available working time per day by the customer demand rate per day (Eq. (4.3)) (Rother & Shook, 2003a, 2003b):

$$Takt\ time = \frac{available\ working\ time\backslash \mathrm{per}\backslash \mathrm{day}}{customer\ demand\ rate\backslash \mathrm{per}\backslash \mathrm{day}} \tag{4.5}$$

In Eq. (4.5), the available working time per demand rate and per day can be realized by regularly releasing a small consistent amount of production through one particular process.

Recently, *heijunka* has gained popularity as a lean tool for smoothing production and improving control of the inventory of finished products (Liker, 2005). The aim of *heijunka* is not only to level the production volume, but also the product mix by using the same sequence of products for each production cycle, thus avoiding peaks and troughs in the production schedule. Demand levelling breaks down the total volume of orders for a given planning period into scheduling intervals (week, day, shift). A *heijunka* calculation then defines a repetitive production sequence (e.g. AAB, AAB, ...) and the lot sizes for that scheduling interval, which dictates the model mix scheduled on a given line.

The interval in which at least one part of each product can be produced must be defined. This interval is called "each part each interval" (EPEI). From a lean management point of view, the shorter the EPEI the better. A short EPEI means that changeover times are reduced and inventory levels can be reduced (Grimaud, Dolgui, & Korytkowski, 2014), as can be seen in Eq. (4.6):

$$\mathrm{EPEI} = \frac{\sum Changeover\ times/product}{Available\ times\ for\ changeovers/period} \tag{4.6}$$

The EPEI represents the time interval in which the sequence "AAB" should be produced. Levelling the production mix means, for example, that instead of assembling all the "Type A" products in the morning and all the "Type B" products in the afternoon, small batches of "A" and "B" are produced alternately. By levelling the production mix and changing between different products more frequently, the supplier can respond to different customer requirements using smaller upstream buffers. The diversity in the model mix requires that any changeover between items has to be extremely simple and quick. *Heijunka* dictates small-lot production sequences, which would be very inefficient without quick setups.

The goals of *heijunka* are to supply customer processes with a constant flow of small lots of different parts and to reduce the bullwhip effect. The first goal contributes to generating a constant demand for parts for upstream processes, reducing or eliminating the need for spare capacity or stocks to cope with peaks of demand. *Heijunka* also attempts to match workloads for each production process and each employee closely to one another. Thus, people can do the same amount of work at the same time and can easily be moved to other production processes. In this way, continuous improvement and flexibility can be attain in line balancing (Coleman & Vaghefi, 1994).

The frame of any *heijunka* implementation should begin with takt time and end with a *heijunka* box. A *heijunka* box (it can also be a wheel or board) is a simple visualization of production using Kanban cards to signal production based on a specified interval of work (e.g. per day). It is used by production staff on the floor and is highly regarded in visualizing processes. The *heijunka* box helps visualize the optimized production schedule. Each square on the board represents a Kanban to signal production of a single product. Each square on a given day represents one product to be made.

The *heijunka* box is a cyclical schedule which is divided into a grid of boxes. The columns represent a specific period of time and the rows represent the product types produced by the subsequent production. This type of organization is in between make-to-stock and make-to-order systems. It is not a make-to-stock system as production is not completely separated from demand and production lots are unequal to orders. It is also not a make-to-order system as demand is fulfilled instantly from the supermarket as long there is a sufficient inventory level. In the case of an inventory shortage, the remaining part of demand is backordered and will be supplemented in the forthcoming period. The assumption is that the entire demand is realized and nothing is lost due to inventory shortages (Grimaud et al., 2014).

4.5.5 Implementation of a *heijunka* Board

First, the *heijunka* board is divided into several sections (single boxes in which a Kanban card can be placed). Each section contains the production orders for one particular hour, from first shift to last shift. The production orders for that 1 h are written on a Kanban card, as illustrated in Fig. 4.25.

Fig. 4.25 Kanban card

KANBAN CARD			
Product	A		
Kanban N°		Upstream	Pre-Assembly line
Products quantity	12	Downstream	Final Assembly line

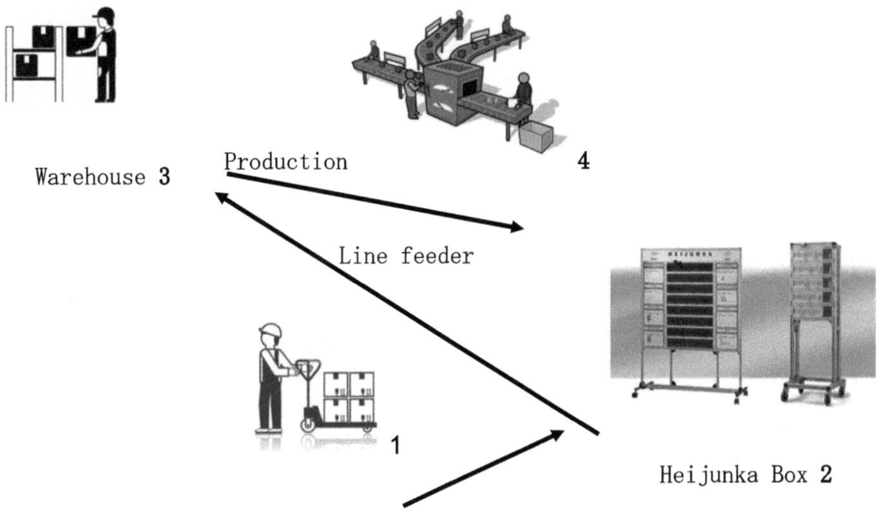

Fig. 4.26 Heijunka box in the production flow

For example, if the aim is to produce X parts each hour, Kanban cards must be placed in each section. The card contains all the information required to produce X parts and works based on a production order. In each cycle, the line feeder picks a card from the box, then also picks all the material required to produce the X parts mentioned on the Kanban card from the warehouse, delivers all materials to the production line and hands over the Kanban card to the line operators. The line operators then have all they need for the next set of parts (X parts in particular) for the next hour (Fig. 4.26).

The *heijunka* board itself acts as a tool for process visualization. In the case of breakdowns, the line will stop operating. This will lead to a delay in cards being picked on the *heijunka* board. If no cards are picked from the board, a deviation from the time slot given by the section and the actual time is verified.

4.5.6 Practical Application of the Kanban System

In this section a practical application of the Kanban system in a company from the automotive industry is set out. The company makes up part of an automotive group that produces the full line of components which go into automotive seating, such as frames, mechanisms and mechatronics, foam pads, trim covers, accessories and electronic and pneumatic systems. The company provides automotive seating with intelligent seat adjustment and customized comfort assembly complete seats, both front and rear, for JIT delivery to its customers' plants.

In the company, a dual-card Kanban system was implemented using P-Kanban and W-Kanban. The P-Kanban operates as a working instruction, being an

P	Line XXXXXXX			
	Internal SKU: 123456789		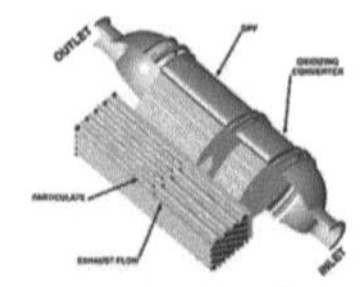	
4	Customer Reference:111111111			
	Designation: Particle filter catalyst			
5		Customer : Line AAAAAA		
6	Quantity/Box	Box	Place	Production flow
7	**40**	Box 600*400*120	Viana	123456789
	Return to supplier			

Fig. 4.27 Kanban card

information transmitter. Once the operator receives a Kanban card, the reference, the quantity to be produced and the type of container to be used are immediately visualized (Fig. 4.27). Normally, to facilitate the work of the operator, for each production line a certain colour is stipulated and for each reference a symbol is given, as shown in Fig. 4.27.

The W-Kanban is a material transfer card, providing information on what material to transfer, in what quantity and where it should be transferred (another production line, shipping area, etc.).

The automotive company in which the Kanban system was implemented has one production line with three stock-keeping units (SKUs) in the following quantities:

- SKU (A)—9 bins
- SKU (B)—5 bins
- SKU (C)—4 bins

For each bin, a W-Kanban will be performed and placed in the following table, arranged in an orderly manner so that material removal from the line is coherent and levelling, eliminating line variability.

It should be noted that when choosing a frame with ten columns, the prime number three is chosen to space the tags sequentially. Thus, we sequentially place the containers in each column respecting the selected spacing, as shown in Fig. 4.28. This distribution is performed by returning to the beginning of the frame until all the containers of the given reference are distributed.

After that, the next three labels are placed, as illustrated in Fig. 4.29.

Then the two final labels are placed on the production levelling board (Fig. 4.30).

Then, we proceed in the same way for the next SKU, taking as a starting point the last container placed on the board, always keeping in mind the selected spacing. Following this procedure, five labels of the SKU (B) were placed on the board (Fig. 4.31).

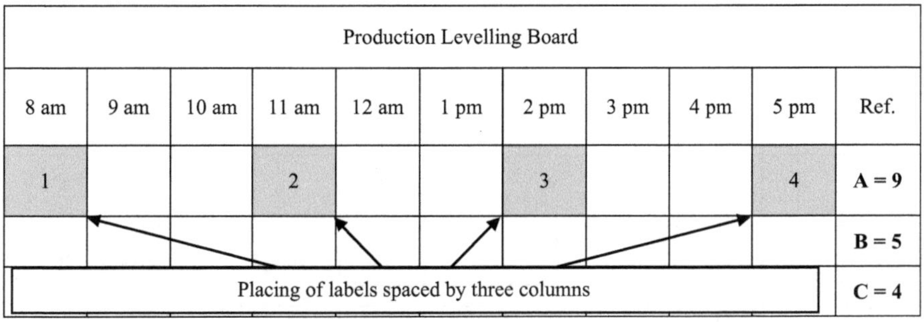

Fig. 4.28 Placement of the four initial labels

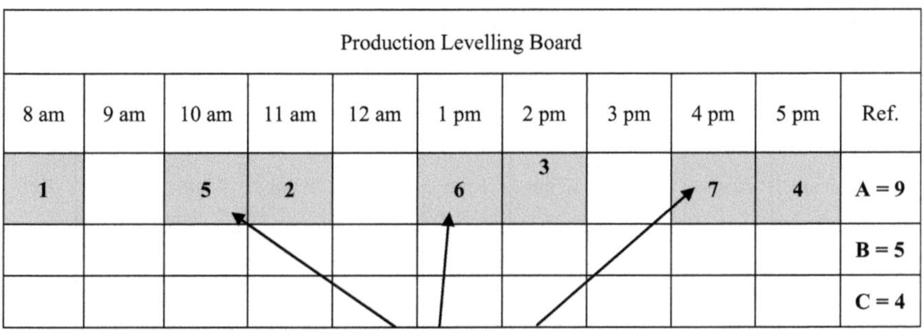

Fig. 4.29 Placement of the next three labels

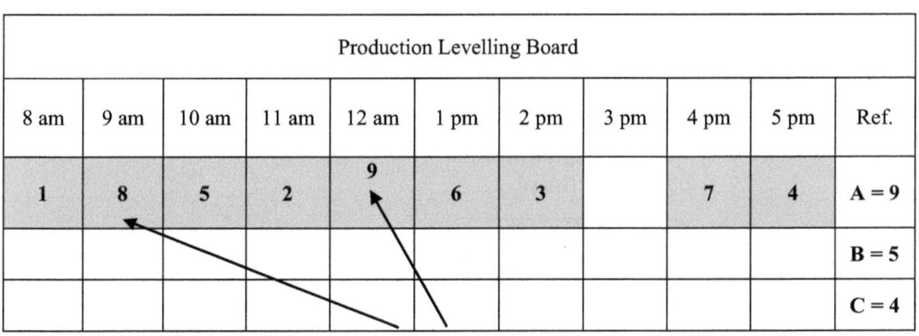

Fig. 4.30 Placement of the two final labels

Finally, the four labels of the last SKU (C) were placed on the production levelling board, as can be seen in Fig. 4.32.

The implementation of the Kanban systems included a supermarket system, which represents the concept of pulling materials from the customer backwards to production (Fig. 4.33). As can be seen in the figure, the final process (customer) goes to the initial process (supermarket) to acquire necessary parts at the time and in the

Production Levelling Board										
8 am	9 am	10 am	11 am	12 am	1 pm	2 pm	3 pm	4 pm	5 pm	Ref.
1	8	5	2	9	6	3		7	4	A = 9
11			12			13	10		14	B = 5
										C = 4

Fig. 4.31 Placement of the five labels of SKU (B)

Production Levelling Board										
8 am	9 am	10 am	11 am	12 am	1 pm	2 pm	3 pm	4 pm	5 pm	Ref.
1	8	5	2	9	6	3		7	4	A = 9
11			12			13	10		14	B = 5
	18	15			16			17		C = 4

Fig. 4.32 Placement of the four labels of SKU (C)

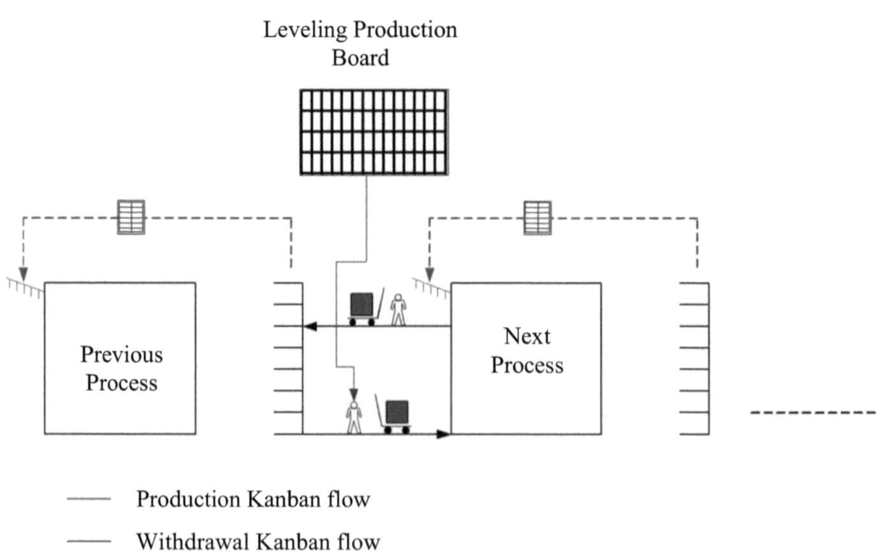

Fig. 4.33 Kanban system: Pull production

quantity that is needed. After that, the initial process immediately produces the quantity just taken to replenish the shelves.

4.5.6.1 Implementation of the Kanban Flow

An important step in the implementation of the Kanban system is the design and management of the Kanban flow. As described above, two types of Kanban cards were implemented in the case study company (P-Kanban and W-Kanban) with the following flow:

1. The supplier of the next process goes to the previous process, provided with a forklift or a conveyor with the number of the W-Kanban necessary to satisfy the requirements, taking the exact number of empty containers withdrawn.
2. As a stock container is removed from the previous process, the P-Kanban that is in the container is also removed and replaced with the W-Kanban.
3. The P-Kanban is then placed on the *heijunka* board. When the number of P-Kanban cards is enough to make a production batch, the P-Kanban labels are grouped and placed on the board near the first-line operator.
4. The P-Kanban thus waits for its turn to be produced, following the order in which it was launched. The line operator has only to look at the *heijunka* board to know which references to produce and in which order.
5. The parts that have been moved by W-Kanban are taken to the next process. On arriving at that process, they are placed on the access ramps to the production line and the W-Kanban is moved and placed on the *heijunka* board, waiting for new levelling.

The sequence begins with the production levelling board, guiding the first processes to be undertaken and their rhythm of withdrawal. Whenever something fails in the process, the failure is quickly noted as part of the whole.

For Monden (1984), if there are any abnormal situations, there is another type of Kanban to deal with such special features or an unforeseen situation. However, such other types of Kanban should only be used as a last resort. The Kanban flow should be kept as close as possible to what has been described above in the Kanban flow.

The other types of Kanban that could be used in abnormal situations are as follows:

- Express Kanbans—Such Kanbans are used to signal the need for more of a particular component so that the manufacturing process does not slow down. They are known also as signal Kanbans. They are used mainly to trigger purchases and to keep the original production conditions running smoothly.
- Emergency Kanbans—As the name suggests, these are used in emergency situations to replace defective parts or to signal a sudden change in the amount of the product that needs to be produced. Emergency Kanbans are used when a part does not work as it is supposed to, or when the conditions of production change.

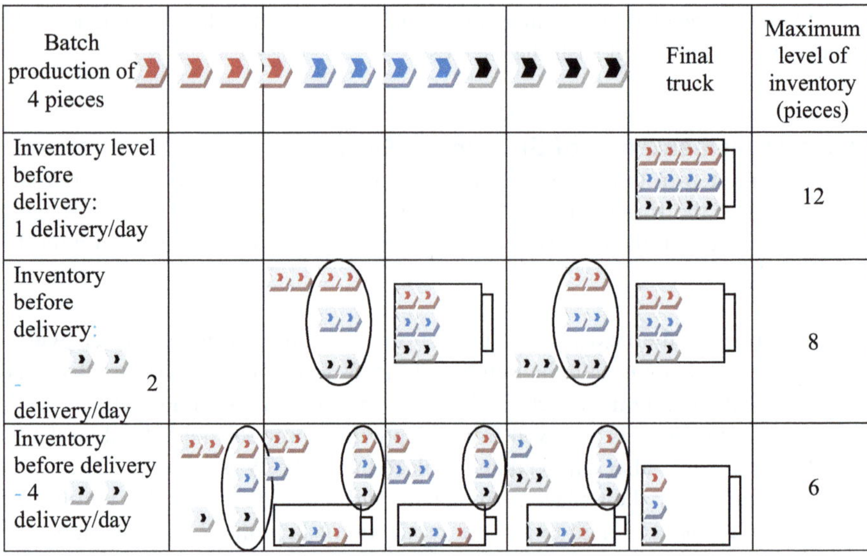

Fig. 4.34 Reduction in inventory level (lot size of four pieces)

- Through Kanbans—These are a combination of W-Kanbans and P-Kanbans and are used when work stations are located side by side in order to speed up production. For instance, if the storage area for components is right next to the place where the product is assembled, having a single Kanban to pull the parts and run them through the production process saves time.

4.5.6.2 Reduction in the Number of Kanbans Versus Stock Reduction

One way of reducing the number of Kanbans in circulation is to ensure interconnection between the SMED methodology and the time taken for the next process to collect containers.

Let us consider the following three cases:

Case 1—Lot size of four pieces and three options for delivery (Fig. 4.34):

- Option 1—1 delivery/day
- Option 2—2 deliveries/day
- Option 3—4 deliveries/day

Case 2—Lot size of two pieces and three options for delivery (Fig. 4.35):

- Option 1—1 delivery/day
- Option 2—2 deliveries/day
- Option 3—4 deliveries/day

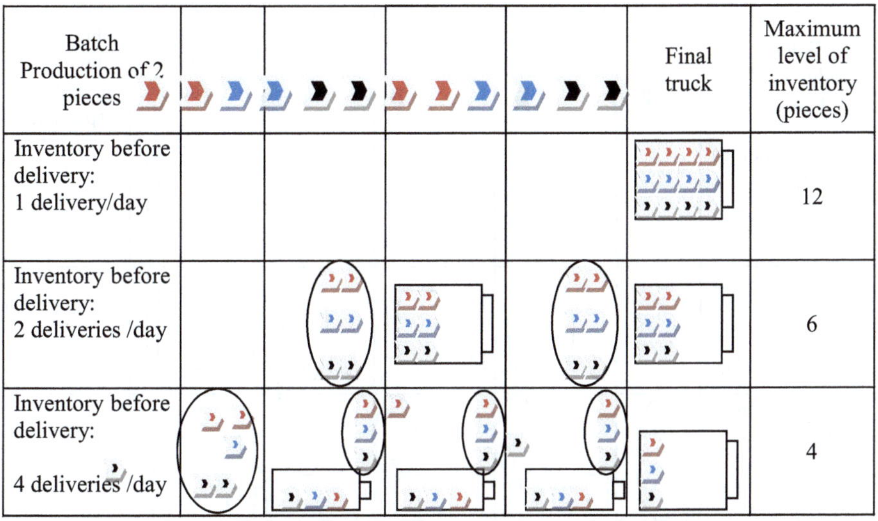

Fig. 4.35 Reduction in inventory level (Lot size: two pieces)

Case 3—Lot size of one piece and three options for delivery (Fig. 4.36):

- Option 1—1 delivery/day
- Option 2—2 deliveries/day
- Option 3—4 deliveries/day

Concerning the three cases illustrated above, as the frequency of delivery increases and batch sizes decrease, the existing level of stocks also decreases (Table 4.15).

4.5.6.3 Results of the Practical Case

The implementation of the Kanban system was carried out in a line of production of components for the automobile industry. This line was considered the test line for the plant, which means that before any methodology could be implemented throughout the factory, it would previously be tested and improved in this line. Only after satisfactory results would the methodologies be implemented in other production lines of the industrial unit. This being a test line, before implementing other methodologies such as 5S and standard work, inter alia, only after all the employees of the line had been trained in these two methodologies was the Kanban implementation carried out. The main characteristics of this test line were as follows:

- Number of SKUs: 6
- Changeover time: 25 min
- Available time: two shifts (900 min)

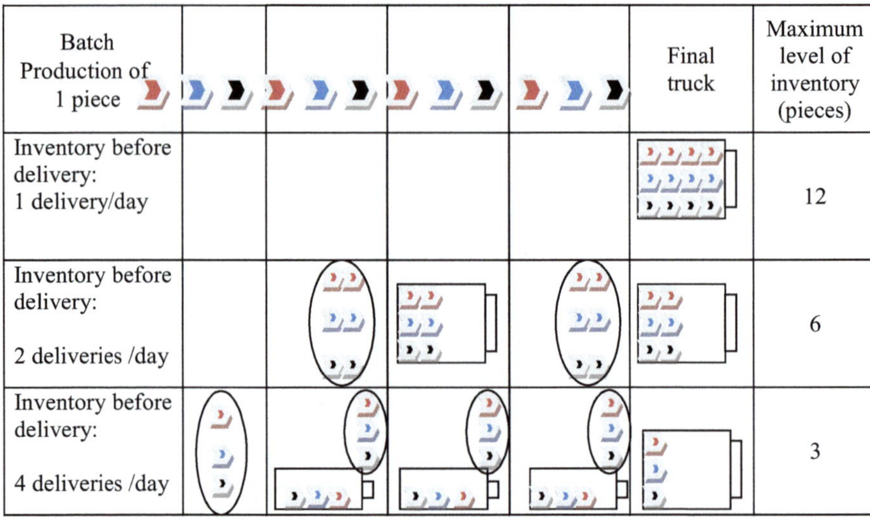

Fig. 4.36 Reduction in inventory level (Lot size: one piece)

Table 4.15 Decrease in stock level (frequency vs. lot size)

	Frequency of delivery			
		One delivery/day	Two deliveries/day	Four deliveries/day
Lot size	Four pieces	12	8	6
	Two pieces	12	6	4
	One piece	12	6	3

- Idle time (faults, lack of material, . . .): 90 min
- Delivery time: 45 min
- Number of employees: six operators + 1 gap leader
- Cycle time: 75 s
- Pieces per container: 24
- Number of daily pieces: 132/reference (SKU)

In Table 4.16 the batch sizes for the reference AAAAA are computed. This method can also be used for the other remaining five SKUs. In the implementation of the Kanban system, first the *heijunka* was implemented to consistently transmit to the production line the same order for 132 pieces by SKU each day. The production levelling board was implemented with the corresponding production levelling labels and the P-Kanban card was used to transmit the production orders to the line.

As can be seen from Table 4.16, the production line only has the capacity to produce four references per day, with a cycle time of around 225 min. Due to the various breaks in the production line, the emergency Kanban was created. This was

Table 4.16 Determination of the initial lot size

Determination of lot sizes			
Line: XXXX	Product: AAAAA		
	Changeover time (CT)=	25	Min
	Available time (AT)=	900	Min
	Suggested changeover time (SCT) = (0.1 × At)	90	Min
	Maximum number of changeovers (MNC) = SCT/CT	4	–
	Projected duration of a series = (AT × 0.9/MNC)	225	Min
	Cycle time (ClT)=	75	Sec
	Minimum lot size = (duration of a series × 60/ClT)	180	Product/Lot
	Lot size to choose (LSC)	192	Product/Lot
	Series real duration	214	Min
	Number of SKUs	6	
	Number of pieces/container (NPC)	24	Pieces/ container
	Number of labels by lot: LSC/NPC	8	Kanban cards
	Number of withdrawn labels = (N° of labels by lot × 2.5)	20	

used regularly as the line was not flexible and there was considerable absenteeism, contributing to breaks in the information flow.

Due to these problems, three working groups were formed: one working group focused on the maintenance of equipment (TPM) to improve its efficiency; the other concentrated on the setup time and the implementation of the lean method (SMED); finally, a logistics group was formed, seeking to achieve the standardization of the work developed by the logistics team, providing material within the expected time and without loss of labels.

After a period of work, the following results were obtained:

- Number of references (SKUS): 6
- Setup time: 15 min
- Production lead time: two shifts (900 min)
- Idle time (faults, lack of material, . . .): 30 min
- Delivery time: 30 min
- Number of employees: six operators + one team leader
- Cycle time: 70 s
- Parts per container: 24 pieces
- Number of daily pieces: 132/reference (SKU)

The determination of the lot sizes is illustrated in Table 4.17.

As can be seen from Table 4.17, the actions performed entailed significant improvements. It became possible to produce six references every day, as well as reducing the level of stocks. The cycle time was reduced to 135 min, which contributed to the production line becoming more flexible and eliminating

Table 4.17 Determination of lot sizes

Determination of lot sizes			
Line: XXXX	Product: AAAAA		
	Changeover time (CT)=	15	Min
	Available time (AT)=	900	Min
	Suggested changeover time (SCT) = (0.1 × At)	90	Min
	Maximum number of changeovers (MNC) = SCT/CT	6	–
	Projected duration of a series = (AT × 0.9/MNC)	135	Min
	Cycle time (CIT)=	70	Sec
	Minimum lot size = (Duration of series × 60/CIT)	116	Product/Lot
	Lot size to choose (LSC)	120	Product/Lot
	Real duration of series	140	Min
	Number of SKUs	6	
	Number of pieces/container (NPC)	24	Pieces/ container
	Number of labels by lot: LSC/NPC	5	Kanban cards
	Number of withdrawn labels = (N° of labels by lot × 2.5)	10	

emergency Kanban. One of the most significant improvements came from logistics, with the standardization of work leading not only to a reduction in the delivery time to 30 min, but more importantly labels no longer disappearing. In addition, the information flows from the truck shipments (customers) throughout the production line and the warehouse of components (suppliers) used only Kanban tags, as in a true pull system.

References

Aggarwal, S. C. (1985). MRP, JIT, OPT, FMS? *Harvard Business Review, 63*(5), 8–16.

Agus, A., & Hajinoor, M. S. (2012). Lean production supply chain management as driver towards enhancing product quality and business performance: Case study of manufacturing companies in Malaysia. *International Journal of Quality and Reliability Management, 29*(1), 92–121.

Akturk, M. S., & Erhun, F. (1999). An overview of design and operational issues of kanban systems. *International Journal of Production Research, 37*(17), 3859–3881.

Archibald, R. (2012). *Project management la gestione di progetti e programmi complessi*. Milan: Franco Angeli.

Ardalan, A. (1997). Analysis of local decision rules in a dual-kanban flow shop. *Decision Science, 28*(1), 195–211.

Ardalan, A., & Diaz, R. (2011). NERJIT: Using net requirement data in Kanban-controlled jumbled-flow shops. *Production and Operations Management, 21*(3), 606–618.

Barkmeyer, E. J. (2007). *An ontology for the e-Kanban business process*. National Institute of Standards and Technology, Internal Report 7404.

Bayo-Moriones, A., Bello-Pintado, A., & Merino-Díaz de Cerio, J. (2010). 5S use in manufacturing plants – Contextual factors and impact on operating performance. *International Journal of Quality & Reliability Management, 27*, 217–230.

Beemsterboer, B., Land, M., & Teunter, R. (2017). Flexible lot sizing in hybrid make-to-order/make-to-stock production planning. *European Journal of Operational Research, 260*, 1014–1023. https://doi.org/10.1016/j.ejor.2017.01.015

Bendavid, Y., Boeck, H., & Philippe, R. (2010). Redesigning the replenishment process of medical supplies in hospitals with RFID. *Business Process Management Journal, 16*(6), 991–1013.

Berkley, B. J. (1992). A review of the Kanban production control research literature. *Production and Operations Management, 1*, 393–411.

Berkley, B. J., & Kiran, A. S. (1991). A simulation study of sequencing rules in a kanban-controlled flow shop. *Decision Science, 22*(3), 559–582.

Blackstone, J. H., Phillips, D. T., & Hogg, G. L. (1982). A state-of-the-art survey of dispatching rules for manufacturing job shop operations. *International Journal of Production Research, 20*(1), 27–45.

Boran, S., & Ekincioğlu, C. (2017). A novel integrated SMED approach for reducing setup time. *International Journal of Advanced Manufacturing Technology, 92*, 3941–3951. https://doi.org/10.1007/s00170-017-0424-9

Braglia, M., Frosolini, M., & Gallo, M. (2017). SMED enhanced with 5-Whys analysis to improve set-upreduction programs: The SWAN approach. *International Journal of Advanced Manufacturing Technology, 90*, 1845–1855. https://doi.org/10.1007/s00170-016-9477-4

Brown, G. D., & O'Rourke, D. (2007). Lean manufacturing comes to China: A case study of its impact on workplace health and safety. *International Journal of Occupational and Environmental Health, 13*(3), 249–257.

Brunet, A. P. (2003). New kaizen in Japan: An empirical study. *International Journal of Operations & Production Management, 33*, 1426–1446.

Carrizo-Moreira, A. (2014). Single minute exchange of die and organizational innovation in seven small and medium-sized firms. In J. G. García-Alcaraz, A. A. Maldonado-Macías, & G. Cortes-Robles (Eds.), *Lean manufacturing in the developing world: Methodology, case studies and trends from Latin America* (pp. 483–499). Cham: Springer.

CDOT. (2017). https://www.codot.gov/business/process-improvement/tools-and-techniques/6s-5s-evaluation-and-implementation-tool

Chalice, R. (2007). *Improving healthcare using Toyota lean production methods: 46 steps for improvement.* Milwaukee, WI: American Society for Quality Press.

Chan, H. K., Yin, S., & Chan, F. T. (2010). Implementing just-in-time philosophy to reverse logistics systems: A review. *International Journal of Production Research, 48*(21), 6293–6313.

Cheung, K. L., Song, J.-S., & Zhang, Y. (2017). Cost reduction through operations reversal. *European Journal of Operational Research, 259*, 100–112. https://doi.org/10.1016/j.ejor.2016.09.052

Chiarini, A. (2013). *Lean organization: From the tools of the Toyota production system to lean office (Perspectives in business culture).* Milan: Springer.

Chiarini, E., & Vagnoni, E. (2015). World–class manufacturing by Fiat. Comparison with Toyota production system from a strategic management, management accounting, operations management and performance measurement dimension. *International Journal of Production Research, 53*, 590–606.

Coleman, B. J., & Vaghefi, M. R. (1994). Heijunka (?): A key to the Toyota production system. *Production and Inventory Management Journal, 35*(4), 31–35.

Davis, W. J., & Stubitz, S. J. (1987). Configuring a kanban system using a discrete optimization of multiple stochastic responses. *International Journal of Production Research, 25*(5), 721–740.

de Salas, K., Lewis, I. J., & Huxley, C. (2017). Using the critical process targeting method to improve SMEs' process understanding: A tale of two Australian case studies. *Business Process Management Journal, 23*, 425–447. https://doi.org/10.1108/BPMJ-06-2014-0052

Dennis, P. (2015). *Lean production simplified* (3rd ed.). New York: CRC Press.

Drickhammer, D. (2005). The Kanban evolution. *Material Handling Management*, 24–26. http://mhlnews.com/technology-amp-automation/kanban-evolution

Emiliani, M. L. (2008). Standardized work for executive leadership. *Leadership and Organization Development Journal, 29*(1), 24–46.

Ferradás, P. G., & Salonitis, K. (2013). Improving changeover time: A tailored SMED approach for welding cells. *Procedia CIRP, Forty Sixth CIRP Conference on Manufacturing Systems, 7*, 598–603. https://doi.org/10.1016/j.procir.2013.06.039

Floyd, R. C. (2010). *Liquid lean: Developing lean culture in the process industries*. Boca Raton, FL: CRC Press.

Fukukawa, T., & Hong, S. C. (1993). The determination of the optimal number of kanbans in a JIT production system. *Computers & Industrial Engineering, 24*(4), 551–559.

Gapp, R., Fisher, R., & Kobayashi, K. (2008). Implementing 5S within a Japanese context: An integrated management system. *Management Decisions, 46*, 565–579.

García-Alcaraz, J. L., & Maldonado-Macías, A. A. (2016). *Just-in-time elements and benefits (Management and industrial engineering)*. Cham: Springer.

Giard, V., & Mendy, G. (2008). Scheduling coordination in a supply chain using advance demand information. *Production Planning and Control, 19*(7), 655–667.

Graupp, P., & Robert, W. (2016). *The TWI workbook: Essential skills for supervisors* (2nd ed.). New York: CRC Press.

Graupp, P., & Wrona, R. (2010). *Implementing TWI: Creating and managing a skills-based culture*. New York: CRC Press.

Grimaud, F., Dolgui, A., & Korytkowski, P. (2014). Exponential smoothing for multi-product lot-sizing with heijunka and varying demand. *Management and Production Engineering Review, 5*(2), 20–26.

Groos, J. M., & Mclnnis, K. R. (2003). *Kanban made simple: Demystifying and applying Toyota's legendary manufacturing process*. Washington DC: AMACOM.

Grünwald, H., Striekwold, P. E. T., & Weeda, P. J. (1989). A framework for qualitative comparison of production control concepts. *International Journal of Production Research, 27*(2), 281–292.

Hadaya, P., & Cassivi, L. (2007). The role of joint collaboration planning actions in a demand-driven supply chain. *Industrial Management and Data Systems, 107*(7), 954–978.

Hall, R. (1983). *Zero inventories*. Homewood, IL: Dow-Jones Irwin.

Ho, S. K. M. (1999). The 5-S auditing. *Managerial Auditing Journal, 14*(6), 294–302.

Holtewert, P., & Bauernhansl, T. (2016). Optimal configuration of manufacturing cells for high flexibility and cost reduction by component substitution. *Procedia CIRP, Research and Innovation in Manufacturing: Key Enabling Technologies for the Factories of the Future – Proceedings of the 48th CIRP Conference on Manufacturing Systems, 41*, 111–116. https://doi.org/10.1016/j.procir.2015.12.042

Huang, C. C., & Kusiak, A. (1996). Overview of kanban systems. *International Journal of Computer Integrated Manufacturing, 9*(3), 169–189.

Huang, P. Y., Rees, L. P., & Taylor, B. W. (1983). A simulation analysis of the Japanese just-in-time technique (with kanbans) for a multiline, multistage production system. *Decision Science, 14*, 326–344.

Imai, M. (1986). *Kaizen: The key to Japan's competitive success*. New York: McGraw-Hill.

Imai, M. (2012). *Gemba kaizen: A commonsense approach to a continuous improvement strategy*. New York, NY: McGraw Hill.

Jackson, T. L. (2009). *5S for healthcare*. New York, NY: Productivity Press.

Karmarkar, U. S. (1986). *Push, pull and hybrid control systems*. Working paper series no. QM8614, Graduate School of Business Administration, University of Rochester.

Kemal Karasu, M., Cakmakci, M., Cakiroglu, M. B., Ayva, E., & Demirel-Ortabas, N. (2014). Improvement of changeover times via Taguchi empowered SMED/case study on injection molding production. *Measurement, 47*, 741–748. https://doi.org/10.1016/j.measurement.2013.09.035

Khamis, N., Abrahman, M. N., Jamaludin, K. R., Ismail, A. R., Ghani, J. A., & Zulkifli, R. (2009). Development of 5S practice checklist for manufacturing industry. *In Proceedings of the World Congress on Engineering, 1*, 978–988.

Kim, T. (1985). Just-in-time manufacturing system: A periodic pull system. *International Journal of Production Research, 23*(3), 553–562.

Kimura, O., & Terada, H. (1981). Design and analysis of pull system: A method of multistage production control. *International Journal of Production Research, 19*, 241–253.

Klipp, P. (2012). *Getting started with Kanban.* https://kanbanery.com/ebook/GettingStartedWithKanban.pdf

Kniberg, H. (2009). *Kanban vs. scrum—how to make the most of both.* http://www.crisp.se/henrik.kniberg/Kanban-vs-Scrum.pdf

Kobayashi, K. (2005). *What is 5S? A content analysis of Japanese management approach.* Unpublished Master's Thesis, Griffith University, Southport.

Kobayashi, K., Fisher, R., & Gapp, R. (2008). Business improvement strategy or useful, tool? Analysis of the application of the 5S concept in Japan, the UK and the US. *Total Quality Management and Business Excellence, 19*, 245–262.

Krishnaiyer, K., & Padmanabhan, P. (2009). *Designing simple electronic Kanban based decision support system – A letter shop case study.* USA: American Society for Quality Conference.

Kumar, V. (2010). JIT based quality management: concepts and implications in Indian context. *International Journal of Engineering Science and Technology, 2*, 40–50.

Kumar, K., & Aouam, T. (2017). Integrated lot sizing and safety stock placement in a network of production facilities. *International Journal of Production Economics, 195*, 74–95. https://doi.org/10.1016/j.ijpe.2017.10.006

Kumar, C. S., & Panneerselvam, R. (2007). Literature review of JIT-KANBAN system. *The International Journal of Advanced Manufacturing Technology, 32*, 393–408.

Lage, J. M., & Godinho Filho, M. (2010). Variations of the Kanban system: Literature review and classification. *International Journal of Production Economics, 125*(1), 13–21.

Lander, E., & Liker, J. (2007). The Toyota production system and art: Making highly customized and creative products the Toyota way. *International Journal of Production Research, 45*(16), 3681–3698.

Lee, L. C., & Seah, K. H. W. (1988). JIT and the effects of varying process and set-up times. *International Journal of Operations and Production Management, 8*(1), 19–35.

LEI. (2008). *Lean lexicon: A graphical glossary for lean thinkers* (4th ed.). Cambridge, MA: Lean Enterprise Institute.

LEI. (2018). Lean Enterprise Institute. https://www.lean.org/

Liker, J. K. (2004). *The Toyota way: 14 management principles from the world's greatest manufacturer* (1st ed.). New York, NY: McGraw-Hill.

Liker, J. K. (2005). *The Toyota way fieldbook.* New York, NY: McGraw-Hill.

Liker, J., & Meier, D. (2006). *The Toyota way fieldbook: A practical guide for implementing Toyota's 4Ps.* New York: McGraw-Hill.

Liker, J. K., & Morgan, J. (2011). Lean product development as a system: A case study of body and stamping development at Ford. *Engineering Management Journal, 23*(1), 16–28.

Linzalone, R., & Schiuma, G. (2015). A review of program and project evaluation models. *Measuring Business Excellence, 19*(3), 90–99.

Lixia, C., & Bo, M. (2011). *How to make 5S as a culture in Chinese enterprises.* Taipei, Taiwan: School of Economics and Management, Changchun University of Science and Technology, Department of Industry Management.

Lockman, L. (2015). Six steps to 5S success. *Industrial Maintenance and Plant Operation, 76*(3), 32–33.

Lozano, J., Saenz-Díez, J. C., Martínez, E., Jiménez, E., & Blanco, J. (2017). Methodology to improve machine changeover performance on food industry based on SMED. *International Journal of Advanced Manufacturing Technology, 90*, 3607–3618. https://doi.org/10.1007/s00170-016-9686-x

Luo, Z., Mo, R., & Ruo, J. (2007). Lean manufacturing and management (2): See through the fog and grasp the essence. *International Manufacturing Technology and Equipment Market, 4*, 101–107.

MacKerron, G., Kumar, M., Kumar, V., & Esain, A. (2014). Supplier replenishment policy using e-Kanban: A framework for successful implementation. *Production Planning and Control: The Management of Operations, 25*(2), 161–175.

Marksberry, P., Rammohan, R., & Vu, D. (2011). A systems study on standardized work: A Toyota perspective. *International Journal of Productivity and Quality Management, 7*(3), 287–303.

Martin, T., & Bell, J. (2011). *New horizons in standardized work: Techniques for manufacturing and business process improvement*. USA: CRC Press.

Mazza, M. M. (2015). Kanban-driven parts feeding within a semi-automated O-shaped assembly line: A case study in the automotive industry. *Assembly Automation, 35*(1), 3–15.

Misiurek, B. (2016). *Standardized work with TWI: Eliminating human errors in production and service processes*. New York: CRC Press.

Miyazaki, S., Ohta, H., & Nishiyama, N. (1988). The optimal operation planning of kanban to minimize the total operation cost. *International Journal of Production Research, 26*(10), 1605–1611.

Monden, Y. (1983). *Toyota production system: Practical approach to production management*. Norcross, GA: Industrial Engineering Management Press.

Monden, Y. (1984). *Toyota production system*. Portland: Engineering and Management Press.

Monden, Y. (1993). *Toyota production system: An integrated approach to just-in time* (2nd ed.). Norcross, GA: Industrial Engineering and Management Press, Institute of Industrial Engineers.

Monden, Y. (1998a). *Toyota production system*. Norcross, GA: Engineering and Management Press.

Monden, Y. (1998b). *Toyota production system: An integrated approach to just-in-time* (3rd ed.). New York: EMP Books.

Monden, Y. (1998c). *Toyota production system: An integrated approach to just-in-time* (3rd ed.). Norcross, GA: Engineering and Management Press.

Nakajima, S. (1988). *Introduction to TPM*. Cambridge, MA: Productivity Press.

Narusawa, T., & Shook, J. (2009). *Kaizen express: Fundamentals for your lean journey*. São Paulo, Brazil: Lean Enterprise Institute.

Nepal, B. P., Yadav, O., & Solanki, R. (2011). Improving the NPD process by applying lean principles: A case study. *Engineering Management Journal, 23*(3), 65–81.

Odgaard, T., Kelly, C., & Laird, J. (2005). *Current practice in project appraisal in Europe*. HEATCO research project (Harmonised European Approaches for Transport Costing and Project Assessment).

Ohno, T. (1988a). *Toyota production system: Beyond large scale production*. New York, NY: Productivity Press.

Ohno, T. (1988b). *Toyota production system: Beyond large-scale production*. Cambridge, MA: Productivity Press.

Ohno, T. (1988c). *Toyota production system: Beyond large-scale production*. Cambridge, MA: Productivity Press.

Onkar, B., & Bhatia, S. (2016). Review on implementation of lean tools and impact of ethnographic factors on 5S. *International Journal for Scientific Research and Development, 4*(06), 717–721.

Osada, T. (1991). *The 5S's: Five keys to a total quality environment*. Tokyo: Asian Productivity Organization.

Owens, D. R. (2006). One good idea: PDCA at the management level. *Quality Progress, 39*(4), 104.

Patchong, A. (2013). *Implementing standardized work: Measuring operators' performance*. Boca Raton, FL: CRC Press.

Phogat, S. (2013). An introduction to applicability of lean in shipbuilding. *International Journal of Latest Research in Science and Technology, 2*(6), 85–89.

Ramdass, K. (2015). Integrating 5S principles with process improvement: A case study. *Portland International Conference on Management of Engineering and Technology*, 1908–1917.

Romano, P., Danese, P., & Bortolotti, T. (2009). The moderating role of JIT links with suppliers on the relationship between lean manufacturing and operational performances. In B. Vallespir & T. Alix (Eds.), *Advances in Production Management Systems. New Challenges, New*

Approaches IFIP WG 5.7 International Conference, APMS 2009, Bordeaux, France, September 21-23, 2009, Revised Selected Papers (pp. 89–96). Berlin, Heidelberg: Springer. https://doi.org/10.1007/978-3-642-16358-6_12

Roser, C. (2017). *"Faster, better, cheaper" in the history of manufacturing: From the Stone Age to lean manufacturing and beyond.* New York: CRC Press.

Rother, M. (2010). *Toyota kata: Managing people for improvement, adaptiveness, and superior results.* New York: McGraw-Hill.

Rother, M., & Shook, J. (2003a). *Learning to see.* Brookline, MA: The Lean Enterprise Institute.

Rother, M., & Shook, J. (2003b). *Learning to see.* Cambridge, MA: Lean Enterprise Institute.

Row, J. R. (2017). *Kanban explained as applied to JIT* (J. Scheid, Ed.). http://www.brighthubpm.com/monitoring-projects/70336-kanban-explained-as-applied-to-jit/

Salonitis, K., & Tsinopoulos, C. (2016). Drivers and barriers of lean implementation in the Greek manufacturing sector. In Factories of the future in the digital environment, Proceedings of the *49th CIRP Conference on Manufacturing Systems, Procedia CIRP, 57*, 189-194.

Schonberger, R. J. (1983). Applications of single-card and dual-card Kanban. *Interfaces, 13*, 56–67.

Sendil Kumar, C., & Panneerselvam, R. (2007). Literature review of JIT-Kanban system. *The International Journal of Advanced Manufacturing Technology, 32*(3–4), 393–408.

Shahabudeen, P., Gopinath, R., & Krishnaiah, K. (2002). Design of bi-criteria kanban system using simulated annealing technique. *Computers & Industrial Engineering, 41*(4), 355–370.

Shapiro, J. (1996). *Monitoring and evaluation.* www.civicus.org/new/media/Monitoring %20and%20Evaluation.pdf

Shingo, S. (1981). *Study of Toyota production system from an industrial engineering viewpoint.* Tokyo: Japan Management Association.

Shingo, S. (1989). *Study of the Toyota production system: From an industrial engineering viewpoint.* New York, NY: Productivity Press.

Shingo, S. (2017a). *Kaizen and the art of creative thinking: The scientific thinking mechanism* (1st ed.). New York: Productivity Press.

Shingo, S. (2017b). *Fundamental principles of lean manufacturing* (1st ed.). Bellingham, WA: Productivity Press.

Singh, B., Garg, S. K., Sharma, S. K., & Grewal, C. (2010). Lean implementation and its benefits to production industry. *International Journal of Lean Six Sigma, 1*(2), 157–168.

Singh, N., & Shek, K.-H. (1990). The development of a Kanban system: A case study. *International Journal of Operations and Production Management, 10*(7), 28–36.

Sipper, D., & Bulfin, R. L., Jr. (1997). *Production: Planning, control, and integration.* New York: McGraw-Hill.

Stefansdottir, B., Grunow, M., & Akkerman, R. (2017). Classifying and modeling setups and cleanings in lot sizing and scheduling. *European Journal of Operational Research, 261*, 849–865. https://doi.org/10.1016/j.ejor.2017.03.023

Sugimori, Y., Kusunoki, K., Cho, F., & Uchikawa, S. (1977). Toyota production system and kanban system materialization of just-in-time and respect-for-human system. *International Journal of Production Research, 15*(6), 553–564.

Sugiura, Y., & Gillespiere, J. K. (2002). *A bilingual handbook on Japanese culture* (2nd ed.). Tokyo: Natsume Books.

Surendra, M. G., Yousef, A. Y., & Ronal, F. P. (1999). Flexible Kanban system. *International Journal of Operations, and Production Management, 19*(10), 1065–1093.

Takahashi, Y., & Osada, T. (1990). *TPM total productive maintenance.* Tokyo: Asian Productivity Organization.

Tapping, D., Kozlowski, S., Archbold, L., & Sperl, T. (2009). *Value stream management for lean healthcare.* Chelsea, MI: MCS Media, Inc.

Toyota Motor Corporation. (1998). *The Toyota production system.* Japan: Toyota Motor Corporation.

Ultriva Inc. (2007). *Manual Kanban vs. electronic Kanban.* http://www.blog.kanban.com/2007/07/manual-kanban-vs-electronic-kanban.html

Vrontis, D., Tardivo, G., Bresciani, S., & Viassone, M. (2016). The competitiveness of the Italian manufacturing industry: An attempt of measurement. *Journal of Knowledge Economics*, 1–17. https://doi.org/10.1007/s13132-016-0397-1

Wan, H. D., & Chen, F. F. (2008). A web-based kanban system for job dispatching, tracking, and performance monitoring. *International Journal of Advanced Manufacturing Technology, 38*(9), 995–1005.

Wang, H., & Wang, H. P. (1991). Optimum number of kanbans between two adjacent workstations in a JIT system. *International Journal of Production Economics, 22*(3), 179–188.

War Production Board, Bureau of Training, Training within Industry Service. (1943). *Job methods: Sessions outline and reference material.* Washington DC: U.S. Government Printing Office.

War Production Board, Bureau of Training, Training within Industry Service. (1944a). *Job instruction: Sessions outline and reference material.* Washington DC: U.S. Government Printing Office.

War Production Board, Bureau of Training, Training within Industry Service. (1944b). *Job relations: Sessions outline and reference material.* Washington DC: U.S. Government Printing Office.

Womak, J., Jones, D., & Roos, D. (1990). *The machine that changed the world.* New York, NY: Rawson Associates.

Zakaria, N., Mohamed, N., Ab Rahid, M., & Rose, A. (2017). Lean manufacturing implementation in reducing waste for electronic assembly line. *MATEC Web of Conferences, 90*, 01048.

Zelinski, P. C. (2005). If 5S is good, try 13S next. *Modern Machine Shop, 77*(9), 12.

Final Considerations Regarding the Just-in-Time Framework

5

5.1 A Framework for the Implementation of Just in Time

Just in time, more than a set of methodologies, is a work and optimization philosophy, which brings organizations to a mental state of waste elimination throughout the value stream chain. By its very nature, it is focused on continuous improvement and waste elimination. The success of JIT is greatly dependent on people's willingness to adopt consistently an improvement culture through the practice of JIT, supported by a set of JIT/lean manufacturing tools.

Grounded on the practical experience acquired during the implementation of a JIT plant and considering the difficulties and suggestions that emerged during the implementation process, this study proposes a framework to serve as a roadmap for JIT implementation in plants. The suggested framework, depicted in Fig. 5.1, seeks to maximize people's involvement and to assure a reliable step-by-step approach, allowing consolidation and the assimilation of knowledge, which are key to the successful implementation of a JIT plant.

This framework aims to work as a guide for the deployment of the several tools described in Chap. 4. Thus, its main purpose is to establish a sequence for putting the tools into practice. The aim is to create an ordering and interlocking of the tools and methodologies supporting JIT that results in efficient and effective outcomes. Based on the experience of this study, one of the greatest difficulties, if not the greatest, in the JIT implementation process in an organization is the ambition to try out and apply as many tools and methodologies as possible, as far as possible. This typically leads to an inconsistent and disorderly deployment of the JIT philosophy, due to the lack of the required time to ensure that the new tool or methodology becomes part of a culture and a habit. The JIT implementation journey typically takes several years until a reasonable level of JIT culture is instilled.

The order in which the various tools are linked is intended to provide a gradual increase in complexity and requirements. Also, in the first stage the primary objective is to work on process stabilization. Only after a certain level of stabilization of processes will the organization be ready to put in practice two important JIT

J. L. Q. Pinto et al., *Just in Time Factory*, Management for Professionals,
https://doi.org/10.1007/978-3-319-77016-1_5

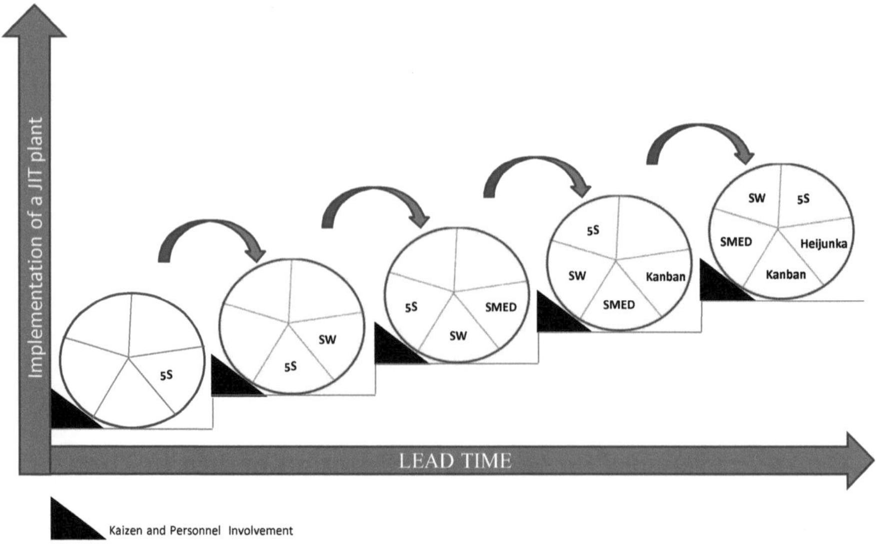

Fig. 5.1 Framework for JIT plant implementation

elements: continuous flow and pull flow. Thus, the objective is to start the application process using 5S, which is simple to understand and apply. The gains achieved with 5S allow employees to see how important it is to apply and develop the JIT concept. When 5S is deeply embedded within the organization, it is the ideal moment to begin the development of a critical spirit and an improvement culture, allowing employees to challenge everything that is around them. At this stage, besides permanent questioning, it is relevant to develop a group spirit, as well as the creativity of the employees. This can be achieved with the introduction of the kaizen philosophy, which goes across all the implementation steps, due to its importance and general applicability. As can be seen in Fig. 5.1, kaizen and personnel involvement go across JIT implementation and should be embedded in all the tools supporting it.

In the second step of the framework, work standardization is approached, following the introduction of order and discipline in the workplace (through the 5S tool) and the implantation of a spirit of continuous improvement (through the kaizen philosophy). At this stage it is necessary to create standardized processes and procedures to support evaluation of continuous improvement, as well as to build consistency and stability in processes.

5S, kaizen and standardized work are particularly associated with process stabilization; thus, after establishing them at an appropriate level of autonomy, we believe that the organization will have the conditions required to evolve to a set of more technical tools, aimed at leading the organization to operate on a continuous takt-based pull flow. For continuous flow, it is essential to work on the reduction of lot sizes, so the third tool to be applied should be SMED. This will create the required conditions for the system to operate on a pull flow, driven by customer needs, at takt time rates. This is achieved through the fourth and fifth tools, Kanban and heijunka respectively. Kanban creates a system that signals supply/replenishment needs throughout the process and heijunka acts on production smoothing by levelling the production volume and mix.

To summarize, we would like to emphasize that there is no defined time range within which JIT implementation should be concluded in a plant or at which each of the tools presented in Fig. 5.1 should be embedded. This should be subject to the assimilation of the concepts by employees and their capability in using the tools. This period can vary significantly according to the extent of requirements, ambition and involvement of an organization in this change process and its actual desire to establish a JIT philosophy. As already mentioned, the employees' involvement is crucial as they are the ones who will support the JIT journey. If this support is not solid, the probability of a tool not being applied and respected is high, with the risk of the JIT system failing.

5.2 Outlook on Just-in-Time Implementation

Just in time, originally a Japanese strategy, has come to be used worldwide and not only in Japanese industry. The objective is to use a number of practices and tools that aim to satisfy the market by producing the products needed, when needed, in the required quantity and at the level of quality required (García-Alcaraz & Maldonado-Macías, 2016). The implementation of JIT can provide many advantages to an enterprise, such as improving profit, reducing costs and inventory and improving quality, among others. These benefits have been demonstrated by many case studies, some of which were presented in Chap. 3.

However, concerning inventories, for example, some authors (e.g. Folinas, Fotiadis, & Coudounaris, 2017) argue that the real effects of modern methods and tools (such as JIT) in terms of inventory management are still unclear. Also, significant differences can be found in various sectors. The same authors found no unanimity when it came to the results in a study in which the objective was to answer the research question: "Do JIT practices affect business performance positively?" The research was focused mostly on manufacturing industries and included 125 companies in northern Greece. Despite some limitations, the results of this study could be representative of the majority of enterprises. The authors suggest that:

- There is no strong link between JIT and zero-level inventory management practices and financial-based performance indicators, even though the manufacturing companies had applied these practices extensively over previous years.
- It was found that simply reducing inventories is in itself not sufficient and a manufacturer must strive for continuous improvement.

- Clients do not wait to purchase a product if it is not immediately available; instead, they will buy a competitor's substitute product that is available. Therefore, all a company's efforts can be eroded if the competitor's product reaches the market first, regardless of how much a firm exerts efforts to compress its inventory levels to reduce costs and improve profits.

In other words, theories that support cost reduction and improved profitability based on the reduction of inventory levels may hold, but sometimes only to a certain extent (Folinas et al., 2017).

The main goal of the JIT production system is to identify hidden problems in the value chain and reduce production system waste while increasing the throughput (Kootanaee, Nagendra Babu, & Talari, 2013). JIT management has a high degree of cultural aspects embedded in its development: it was first applied in Japanese manufacturing organizations and is related to Japanese cultural characteristics. Thus, the application of the same model in different cultures may not be fitting. This aspect was perhaps the main reason for several unsuccessful applications outside Japan.

As mentioned above and drawing on other authors, García-Alcaraz and Maldonado-Macías (2016) identified a total of 34 elements that ensure successful JIT implementation. They divided these factors into three groups: elements associated with human resources; elements associated with the production process; elements associated with the product. Thus, it is important to verify all of these factors to attain successful JIT production. There are several instances of successful JIT implementation outside Japanese culture. For example Kootanaee et al. (2013) present several benefits of JIT implementation and some limitations. They also address the prerequisites for a JIT economic programme. However, action and preparation are required in the organization prior to JIT programme implementation, such as plant evaluation, management influence, housekeeping activities and organizational flexibility; these elements are essentially associated with human resources, the production process and the product.

In the current economic environment, organizations are forced to move at high speed because of the differentiated competition originating from all quarters. It is therefore necessary for all support mechanisms to be activated and entirely operationalized so that challenges do not become dangerous. These support mechanisms are none other than the various tools described and used in this study. The verification of their existence and activation is related to the implementation stage and analysis of fulfilment. It should be noted that the organization's improvement is continuous. There is always something that can yet be improved whenever there are interventions among people and equipment. However, improvement will only be manifested if people take the time to intervene on the shop floor, to observe and wish to improve.

As remarked on in Sect. 5.1, JIT is more than a set of methodologies. It is an attitude and a way of thinking. The wholehearted involvement of employees is one of the key aspects in assuring success when implementing JIT. However, the systematic organization of tools is also essential in ensuring an efficient model.

Nonetheless, these aspects alone are not sufficient. All elements should be properly and sequentially implemented in such a way that knowledge consolidation and assimilation are achieved, these being key factors for the success of the model.

In our case study, we found that it is not recommended to try all tools as soon as possible. Thus, the way in which the tools are conveyed and applied is decisive. Indeed, an organization should only progress to the next stage when the previous stage is completely assimilated, meaning that there is autonomy in implementing and developing the tools.

In the implementation of JIT, it is necessary to undertake the following steps:

- Identify the purpose of the implementation of JIT methodology.
- Analyse the importance of the lean production concept within the framework of a JIT plant.
- Determine the importance of various tools in the JIT implementation, guided by its methodology.
- Verify the existence of improvements and at what levels when implementing this methodology in an organization.

The purpose of JIT implementation is waste elimination. Thus, an enterprise should:

- Carry out only such actions that create added value;
- Ensure all activities performed create value from the client's point of view;
- Aim for a flow within all the necessary steps in the project implementation with maximum value added;
- Produce only what is requested by the client;
- Aim for continuous and efficient waste reduction.

Successful JIT implementation aims to achieve more with less—less time, fewer stocks, fewer lower quality products, less manpower, less space—throughout the process, the least possible.

This work has verified that the application of a framework for a JIT plant requires a lean production approach. This concept leads to the identification and elimination of waste based on the continuous improvement method, following the product from the order to delivery to the client. With this constant focus on the concept of eliminating waste, it has been possible to demonstrate that the achievement of a JIT plant is related to lean production. This evidence has been verified for all tools, the ultimate objective of which is to optimize processes and production flows.

The case study highlights the need to examine the various tools of the JIT model employed, allowing the solidification of the proposed model. In this regard, the following remarks are pertinent:

- The application of 5S is not exclusive to those involved in production, but pertains to all collaborators. Thus, the methodology should be applied in all departments, such as management, logistics, maintenance, etc.

- In the development phase it is necessary to stimulate responsibility on the part of employees. It is essential to create among employees a spirit of continuous improvement. Their imaginations should be tested in a simple but systematic manner, analysing inefficient processes and proposing solutions. It should never be assumed that the solution appears only by replacement and by using the kaizen methodology. "Do it yourself" is the spirit to be created and developed. In turn, in the standardized work phase, each employee must be aware of his/her role; employees need to know what to do, how to do it well and when and how many times to carry out a certain task.
- At the execution level—and after employees exhibit improved spirit—more specific tools (heijunka, SMED and Kanban) should/can be more strongly implemented. Here it is necessary to exert great effort and dedication so that the results become real and sustainable and ensure that employees have an active and critical voice.
- Verification of improvements when implementing JIT, as well as at what levels, should be undertaken for all the tools as carefully as possible.

5.3 Final Remarks

JIT can be successful if there is knowledge of the various tools to be implemented. Rigour is extremely important, especially for the JIT implementation study phase until the solidification phase. With JIT implementation, an organization evolves towards acquiring a greater sensitivity to the waste that occurs during the tasks achieved. The acquisition of this sensitivity can be translated into the application of kaizen and always wanting to do better. This results in substantial improvements in working conditions, leading to the organization operating in a lean production mode. Wanting to do ever more with less becomes not a step but a goal to be achieved. In the application of this methodology, one of the most visible benefits is the possibility of looking at an organization in its entirety, from the client to the supplier, allowing the study of the optimization of the whole chain. For this, the involvement of all departments is essential.

However, it is also necessary here to make some recommendations. These are focused on two tools: standardization of work and the production Kanban (P-Kanban). All the others have been properly studied and implemented in various publications. Regarding standardized work, the model used in the case study is quite extensive and its implementation faces certain difficulties in the everyday activities of an organization as the time available to implement something that can be complex is limited. Thus, this model should be optimized by simplifying its application and understanding. With respect to the P-Kanban, it has been possible to verify that this is an important tool for an organization that intends to operate according to the JIT model. However, one of the objectives of the JIT model is to eliminate waste and Kanban is still one of the factors that creates waste, although it is one of the tools that most contributes to its reduction. It should be noted that Kanban is one of the most desirable analytical tools. Hence, increasingly organizations are beginning to emerge that implement other types of Kanban, such as constant work-in-progress (CONWIP), paired-cell overlapping loops of cards with authorization (POLKA) and control of balance by card-based navigation (COBACABANA) (Thürer, Stevenson, & Protzman, 2016). For example, according to Leonardo, Sereno, Silva, Sampaio, Massote, and Simões (2017), the CONWIP Kanban has advantages over the traditional method used in our case study.

Finally, the study can be concluded with a very useful slogan for the development and implementation of the JIT methodology in an organization: "The example is not just one, but the only way to teach" (unknown author).

References

Folinas, D. K., Fotiadis, T. A., & Coudounaris, D. N. (2017). Just-in-time theory: The panacea to the business success? *International Journal of Value Chain Management, 8*, 171–190. https://doi.org/10.1504/IJVCM.2017.085485

García-Alcaraz, J. L., & Maldonado-Macías, A. A. (2016). *Just-in-time elements and benefits, management and industrial engineering*. Cham: Springer.

Kootanaee, A. J., Nagendra Babu, K., & Talari, H. F. (2013). Just-in-time manufacturing system: From introduction to implement. *International Journal of Economics, Business and Finance, 1*(2), 07–25.

Leonardo, D. G., Sereno, B., Silva, D. S. A., Sampaio, M., Massote, M. A. A., & Simões, J. C. (2017). Implementation of hybrid Kanban-CONWIP system: A case study. *Journal of Manufacturing Technology Management, 28*(6), 714–736. https://doi.org/10.1108/JMTM-03-2016-0043

Thürer, M., Stevenson, M., & Protzman, C. W. (2016). Card-based production control: A review of the control mechanisms underpinning Kanban, ConWIP, POLCA and COBACABANA systems. *Production Planning and Control, 27*(14), 1143–1157. https://doi.org/10.1080/09537287.2016.1188224

MIX
Papier aus verantwortungsvollen Quellen
Paper from responsible sources
FSC® C105338

If you have any concerns about our products,
you can contact us on
ProductSafety@springernature.com

In case Publisher is established outside the EU,
the EU authorized representative is:
Springer Nature Customer Service Center GmbH
Europaplatz 3, 69115 Heidelberg, Germany

Printed by Libri Plureos GmbH
in Hamburg, Germany